W9-AOG-992

WITHDRAWN
L. R. COLLEGE LIBRARY

QH
485
.A9
1982

124835

DATE DUE

| | | | |
|---|---|---|---|
| | | | |
| | | | |
| | | | |
| | | | |
| | | | |
| | | | |
| | | | |
| | | | |
| | | | |
| | | | |
| | | | |
| | | | |

CAMBRIDGE & YORK
CAMBRIDGE UNIVERSITY STORES

SECOND EDITION **Reproduction in mammals**

BOOK **1** *Germ cells and fertilization*

EDITED BY  C. R. AUSTIN
*Formerly Fellow of Fitzwilliam College*
*Emeritus Charles Darwin Professor of Animal Embryology*
*University of Cambridge*

R. V. SHORT, FRS
*Professor of Reproductive Biology*
*Monash University, Australia*

DRAWINGS BY JOHN R. FULLER

CARL A. RUDISILL LIBRARY
LENOIR RHYNE COLLEGE

Cambridge University Press
*Cambridge*
*London   New York   New Rochelle*
*Melbourne   Sydney*

*QH*
*485*
*.A9*
*1982*

*124835*
*may 1983*

Published by the Press Syndicate of the University of Cambridge
The Pitt Building, Trumpington Street, Cambridge CB2 1RP
32 East 57th Street, New York, NY 10022, USA
296 Beaconsfield Parade, Middle Park, Melbourne 3206, Australia

© Cambridge University Press 1972, 1982

First published 1972
Reprinted 1973, 1978
Second edition 1982

Printed in Great Britain at the University Press, Cambridge

Library of Congress catalogue card number: 81-18060

*British Library Cataloguing in Publication Data*

Reproduction in mammals. – 2nd edn.
Book 1: Germ cells and fertilization

1. Mammals – Reproduction
I. Austin, C. R.   II. Short, R. V.
III. Fuller, John R.
599.01'6   QL739.2
ISBN 0 521 24628 8 hard covers
ISBN 0 521 28861 4 paperback
(First edition:
ISBN 0 521 08408 3 hard covers
ISBN 0 521 09690 1 paperback)

# CONTENTS

# CONTRIBUTORS TO BOOK 1

C. R. Austin
Manor Farm House
Toft
Cambridge CB3 7RY
UK

T. G. Baker
School of Medical Sciences
Bradford
West Yorkshire BD7 1DP
UK

J. M. Bedford
Cornell Medical Center
525 East 68th Street
New York, NY 10021
USA

A. G. Byskov
The Finsen Laboratory
2100 København Ø
Strandboulevarden 49
Denmark

M. J. K. Harper
Medical School
7703 Floyd Curl Drive
San Antonio, Texas 78284
USA

B. P. Setchell
Department of Animal Sciences
Waite Agricultural Research Institute
Glen Osmond, SA
Australia 5064

# *PREFACE TO THE SECOND EDITION*

In this, our Second Edition of *Reproduction in Mammals*, we are responding to numerous requests for a more up-to-date and rather more detailed treatment of the subject. The First Edition was accorded an excellent reception, but the Books 1 to 5 were written ten years ago and inevitably there have been advances on many fronts since then. As before, the manner of presentation is intended to make the subject matter interesting to read and readily comprehensible to undergraduates in the biological sciences, and yet with sufficient depth to provide a valued source of information to graduates engaged in both teaching and research. Our authors have been selected from among the best known in their respective fields.

Book 1 deals with the origin of the germ cells, the formation of the gonads, the production and properties of the gametes, gamete transport, and the events leading up to and including fertilization.

*From the Preface to the First Edition*

*Reproduction in Mammals* is intended to meet the needs of undergraduates reading Zoology, Biology, Physiology, Medicine, Veterinary Science and Agriculture, and as a source of information for advanced students and research workers. It is published as a series of eight small textbooks dealing with all major aspects of mammalian reproduction. Each of the component books is designed to cover independently fairly distinct subdivisions of the subject, so that readers can select texts relevant to their particular interests and needs, if reluctant to purchase the whole work. The contents lists of all the books are set out on the next page.

# BOOKS IN THE FIRST EDITION

**Book 1. Germ cells and fertilization**
Primordial germ cells *T. G. Baker*
Oogenesis and ovulation *T. G. Baker*
Spermatogenesis and the spermatozoa *V. Monesi*
Cycles and seasons *R. M. F. S. Sadleir*
Fertilization *C. R. Austin*
**Book 2. Embryonic and fetal development**
The embryo *A. McLaren*
Sex determination and differentiation *R. V. Short*
The fetus and birth *G. C. Liggins*
Manipulation of development *R. L. Gardner*
Pregnancy losses and birth defects *C. R. Austin*
**Book 3. Hormones in reproduction**
Reproductive hormones *D. T. Baird*
The hypothalamus *B. A. Cross*
Role of hormones in sex cycles *R. V. Short*
Role of hormones in pregnancy *R. B. Heap*
Lactation and its hormonal control *A. T. Cowie*
**Book 4. Reproductive patterns**
Species differences *R. V. Short*
Behavioural patterns *J. Herbert*
Environmental effects *R. M. F. S. Sadleir*
Immunological influences *R. G. Edwards*
Aging and reproduction *C. E. Adams*
**Book 5. Artificial control of reproduction**
Increasing reproductive potential in farm animals *C. Polge*
Limiting human reproductive potential *D. M. Potts*
Chemical methods of male contraception *H. Jackson*
Control of human development *R. G. Edwards*
Reproduction and human society *R. V. Short*
The ethics of manipulating reproduction in man *C. R. Austin*
**Book 6. The evolution of reproduction**
The development of sexual reproduction *S. Ohno*
Evolution of viviparity in mammals *G. B. Sharman*
Selection for reproductive success *P. A. Jewell*
The origin of species *R. V. Short*
Specialization of gametes *C. R. Austin*
**Book 7. Mechanisms of hormone action**
Releasing hormones *H. M. Fraser*
Pituitary and placental hormones *J. Dorrington*
Prostaglandins *J. R. G. Challis*
The androgens *W. I. P. Mainwaring*
The oestrogens *E. V. Jensen*
Progesterone *H. B. Heap and A. P. F. Flint*
**Book 8. Human sexuality**
The origins of human sexuality *R. V. Short*
Human sexual behaviour *J. Bancroft*
Variant forms of human sexual behaviour *R. Green*
Patterns of sexual behaviour in contemporary society *M. Schofield*
Constraints on sexual behaviour *C. R. Austin*
A perennial morality *G. R. Dunstan*

viii

*BOOKS IN THE SECOND EDITION*

**Book 1. Germ cells and fertilization**
Primordial germ cells and regulation of meiosis *A. G. Byskov*
Oogenesis and ovulation *T. G. Baker*
The egg *C. R. Austin*
Spermatogenesis and spermatozoa *B. P. Setchell*
Sperm and egg transport *M. J. K. Harper*
Fertilization *J. M. Bedford*
**Book 2. Embryonic and fetal development**
The embryo *A. McLaren*
Implantation and placentation *M. B. Renfree*
Sex determination and differentiation *R. V. Short*
Fetal development and birth *G. C. Liggins*
Pregnancy losses and birth defects *P. A. Jacobs*
Experimental embryology *R. L. Gardner*
**Book 3. Hormonal control of reproduction**
The hypothalamus and the anterior pituitary *F. J. Karsch*
The posterior pituitary *D. W. Lincoln*
The pineal gland *G. A. Lincoln*
The testis *D. de Kretser*
The ovary *D. T. Baird*
Oestrous and menstrual cycles *R. V. Short*
Pregnancy *R. B. Heap and A. P. F. Flint*
Lactation *A. T. Cowie*
**Book 4. Reproductive fitness**
Reproductive strategies *R. M. May and D. Rubenstein*
Species differences in reproductive mechanisms *R. V. Short*
Genetic control of fertility *R. B. Land*
Effects of the environment on reproduction *B. K. Follett*
Sexual behaviour *E. B. Keverne*
Immunoreproduction *N. Alexander*
Reproductive senescence *C. E. Adams*
**Book 5. Manipulating reproduction**
Increasing productivity in farm animals *K. Betteridge*
Today's and tomorrow's contraceptives *R. V. Short*
Contraceptive needs of the developing world *D. M. Potts*
Risks and benefits of contraception *M. P. Vessey*
Augmenting human fertility *D. T. Baird*
Our reproductive options *A. McLaren*
Barriers to population control *C. R. Austin*

# 1

# Primordial germ cells and regulation of meiosis

*ANNE GRETE BYSKOV*

Reproduction in mammals is a remarkably complex process with its own systems of behavioural and physiological events, but in the end it all depends on the fusion of specialized germ cells from the male and the female. At the time of fertilization, the germ cells, or gametes, are haploid in chromosome number and DNA content, a state that is achieved through cell divisions of an unusual kind making up the process of meiosis. Meiosis consists of two consecutive divisions: during the first, both the chromosome number and the DNA content are halved; in the second, only the DNA content is halved. Before the germ cells enter meiosis they pass through a phase of DNA synthesis, just as other cells do in the S phase of the cell cycle. Before entering the first meiotic prophase, the germ cells are therefore equipped with a DNA quota of 4n. After the two meiotic divisions the haploid germ cells are only 1n in terms of DNA content. The meiotic process is unique, not only because it results in the formation of haploid cells, but also because the homologous chromosomes during the first meiotic prophase form pairs and exchange DNA. Since the homologous chromosomes derive from the male and the female, their genes will be mixed by this exchange. The haploid germ cells therefore contain genes from both maternal and paternal sources.

Before meiosis, the germ cells exist as diploid cells, and have been traced back to early stages of embryonic development, at which time they are known as primordial germ cells. The earliest stage at which they have been recognized in man is that of the 24-day embryo, when they occupy a site outside the embryo itself, at the top of the yolk sac. At this time they have just acquired stainability for alkaline phosphatase, so that they can be picked out from other cells of the embryo. Where their non-staining precursors existed before this is still a mystery. From that site, as the embryo develops, the primordial germ cells migrate through the tissues to the gonadal primordia, the structures from which ovary and testis will form. The story is a fascinating one, and it is being added to every year, but there are still many gaps in the narrative.

### Life history of the germ cells
The haploid germ cells are produced in the adult gonad, but the diploid germ-cell line is established during early embryogenesis. Before or shortly

1

after the egg starts to cleave, a certain region of the cytoplasm, the germ plasm, is set aside and will determine the next generation of germ cells. This theory was proposed near the beginning of this century by August Weismann, who worked with the nematode *Ascaris*. That a specific part of the egg cytoplasm will give rise to the new germ cells has later been confirmed in many animals.

In the fruit fly *Drosophila* the origin of the germ plasm can be traced back to the earliest cleavage stage of the embryo. The fertilized fly oocyte contains a granular cytoplasm at its posterior end, the so-called germinal cytoplasm or polar plasm. When the egg cleaves, some cells at the posterior end (the polar cells) will contain this granular polar plasm, and so will become germ cells. Just before the polar cells are formed, the polar granules are fragmented and the smaller granules associate with ribosomes. It has been proposed that the granules contain messenger RNA to code for a specific protein synthesis that determines, or is necessary for, germ cell differentiation. A similar germinal cytoplasm containing granules is present in the frog embryo. In the young rat and mouse embryos a dense filamentous material, 'nuage', which appears to consist of protein and RNA, might be the determinant of the germ cells in these species, since a similar material is confined to the primordial germ cells.

The germ cell is unipotent, since it can only differentiate in one direction, either into a spermatozoon or an egg, whereas the fusion product of the germ cells, i.e. the fertilized egg, is totipotent and gives rise to all the somatic cells as well as the new germ-cell line.

## Origin, migration and multiplication of germ cells

When the gonads become morphologically differentiated into testis or ovary, the germ cells of the male are initially termed prospermatogonia, and then spermatogonia; those of the female are oogonia. These cells divide mitotically until meiosis is initiated (Fig. 1.1). After beginning the meiotic process, germ cells can no longer divide mitotically, and they are therefore incapable of increasing their number. This is crucial for the female mammal in most species since all her germ cells will be transformed to oocytes during early stages of development – in many instances during fetal life. Thus, the female starts her fertile life with a finite number of germ cells – just under half a million for a woman (see Fig. 2.9). Numerous studies have shown that the killing of small oocytes early in life invariably leads to sterility. No new oocytes can arise after meiosis has begun and all the oogonia are transformed to oocytes. In some rare exceptions, as in the primitive primates, the lemurs from Madagascar, isolated groups of oogonia persist in the adult's ovaries. These germ cells can incorporate thymidine, indicating that they may have retained the ability to multiply. It is doubtful, however, whether these isolated primordial germ cells will ever enter meiosis; they may rather be considered as embryonic remnants.

In sharp contrast to the female pattern, the male retains a stem cell population of mitotically dividing spermatogonia usually throughout adulthood. During his fertile life, spermatogonia continue to enter meiosis and be transformed into spermatocytes and spermatozoa.

Until only a few years ago the origin of the primordial germ cells was still disputed. The most generally accepted theory stated that the germ cells originated in the coelomic epithelium covering the gonads. This epithelium was therefore termed the 'germinal epithelium', a term that is highly misleading: there is no evidence to support the idea that germ cells arise from this epithelium. On the contrary, several experiments have shown that the source of the primordial germ cells is not only extragonadal but even extraembryonic.

Germ cells derive from primordial cells that are first seen outside the embryo proper, in the epithelium of the dorsal endoderm of the yolk sac, near the developing allantois (Fig. 1.2). They migrate along the hindgut of the embryo until they reach the tissue covering the ventral area of the primitive kidney, the mesonephros, where the gonad develops. This tissue

Fig. 1.1. Life cycles of male and female germ cells. Germ cells of both sexes divide mitotically until or shortly after gonadal sex differentiation. The female germ cells all enter meiosis in fetal life, whereas the male germ cells keep a resting stem cell population which can divide mitotically and from which meiotic cells continue to emerge throughout adult life.

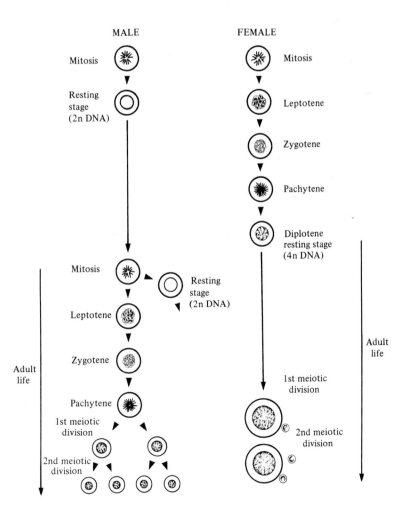

constitutes the genital ridge which will later form an ovary or a testis, and when the germ cells arrive here they come to be called gonia (oogonia or prospermatogonia). Migrating primordial germ cells can be recognized from most of the somatic cells by several criteria. Their nuclei are spherical with little heterochromatin except in their large sponge-like nucleoli. Their cytoplasm is characterized by a paucity of organelles, which gives the cells a clear appearance. In many species these germ cells can also be traced by their high content of cytoplasmic alkaline phosphatase and certain esterases, and sometimes by their content of glycogen.

The reaction for alkaline phosphatase has been an extremely useful tool in tracing and counting the migrating primordial germ cells and pro-liferating gonia. That these positively reacting cells are indeed germ cells is borne out primarily by their entrance into meiosis early in life in the female. After the onset of meiosis the germ cells soon lose their positive reaction for alkaline phosphatase. Other evidence to support the notion that the alkaline phosphatase activity identifies the primordial germ cells during early developmental stages derives from different studies. In certain mouse strains bearing the $W$ gene, the homozygote $WW$ is sterile because of ovarian atrophy and lack of follicular development, while the homozygote $ww$ and the heterozygote $Ww$ are fertile. In the developing embryos of the sterile $WW$ mutant only a few cells with a positive alkaline phosphatase reaction can be traced. A similar picture is seen in irradiated mouse fetuses. Germ cells, and in particular primordial germ cells and small oocytes, are extremely sensitive to irradiation. Shortly after irradiation of the pregnant

Fig. 1.2. The apparent site of origin of primordial germ cells, as demonstrated in a human embryo at 24 days of gestation. (After Fig. 1, Plate 1, in E. Witschi. *Contrib. Embryol.* **32**, 69 (1948).)

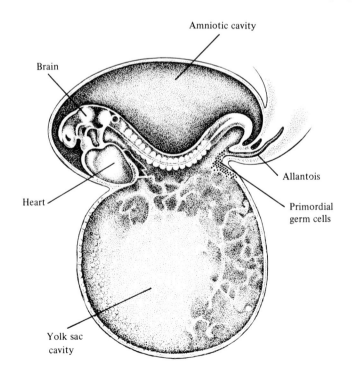

mouse, her fetuses contain fewer cells with the positive enzyme reaction. With higher doses of irradiation a correspondingly smaller number of these cells remains. This again results in fewer or no germ cells in the adult gonad. It thus appears that the cells that show the positive alkaline phosphatase reaction during early embryonic and fetal life are indeed the germ cells.

In mammals the primordial germ cells travel through the tissues by active amoeboid movements, which have been demonstrated by time-lapse photography. They probably find their way from the extraembryonic sites to the presumptive gonads by following chemotactic attractive forces. In culture systems of avian tissues, the epithelial layer of the developing gonads can be shown to attract undifferentiated germ cells. This attraction has a cellular specificity for germ cells, which is species non-specific: that is to say, germ cells from the hindguts of mouse embryos transplanted into the coelom of 2.5-day-old chick embryos migrate towards the developing gonad. We can infer that the attraction of germ cells toward the gonadal area is attributable to a diffusible substance, but the positive demonstration of such a substance and identification of its chemical composition have still to be achieved.

During migration, germ cells sometimes fail to reach the gonadal ridges, and in almost all cases they soon die. However, sometimes they become isolated and survive in the foreign tissues. For unknown reasons such extragonadal primordial germ cells may occasionally start to proliferate and result in the formation of germ-cell tumours.

Only about 100 germ cells are present at the beginning of the migration, but their number then increases rapidly as the cells divide by mitosis. In the mouse the number of germ cells increases from 100 on day 8 of fetal life to about 5000 4 days later. In the human fetus a few hundred are present 3 weeks after conception, but in the fifth month the number has increased to reach a peak of 7 million (Fig. 1.3, and Table 1.1). The enormous increase in germ cell number in the ovary is the result of vigorous mitotic activity. In the testis, too, the spermatogonia divide mitotically and increase in number.

In the human ovary the transformation of oogonia to oocytes by entrance into meiotic prophase is initiated from the third month of gestation. During the next four to five months, more and more oogonia enter meiosis, and by the time of birth or soon after, all female germ cells are oocytes. A dramatic fall in the number of germ cells during the second half of fetal life is seen in both sexes, and seems to be correlated with the period when meiosis is in progress in the female. Closer examination reveals that many oocytes degenerate in the zygotene and pachytene stages. Possibly the crossing-over that takes place between homologous chromosomes is a high-risk procedure, resulting in chromosomal breaks and other errors, so that few oocytes survive for further development.

The reason why so many male germ cells degenerate during the same period is more uncertain. The evidence we have indicates that the male

Table 1.1. *Ages at which germ cells, gonads and sex ducts develop in the human fetus*

| Age of embryo | Morphological event |
| --- | --- |
| 24 days | First primordial germ cells recognizable |
| During 4th week | Mesonephric tubules form |
| 25–30 days | Primordia of Wolffian ducts form |
| 31–35 days | Primordia of sex glands appear as thickening of coelomic epithelium |
| 28–35 days | Migration of germ cells to gonadal primordia |
| 44–48 days | Mullerian duct appears |

*Source*: From H. Peters and K. P. McNatty. *The Ovary*. Paul Elek; London, Toronto, Sydney and New York (1980).

Fig. 1.3. Graph showing numbers of germ cells in the ovaries of fetuses and neonates in man, monkey, rat and guinea pig. (From T. G. Baker, chapter 10, in *Reproductive Biology*, ed. H. Balin and S. R. Glasser. Excerpta Medica; Amsterdam (1972).)

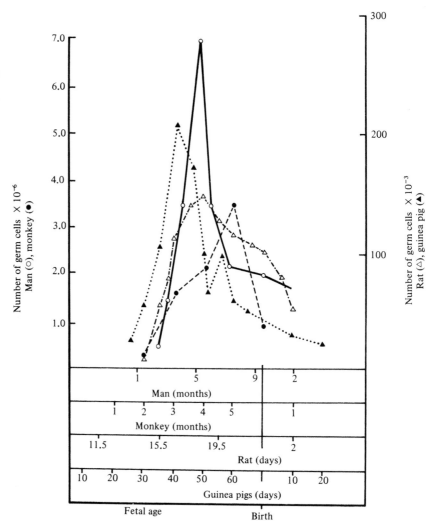

germ cells, simultaneously with the female ones, become exposed to stimuli that are responsible for the induction of meiosis. The cells are, however, prevented from advancing through the meiotic stages by yet other agents. These influences opposing meiosis may cause degeneration of those male germ cells that have responded to the inducing stimuli.

### The undifferentiated gonad

When the germ cells reach the area of the presumptive gonad (the genital or gonadal ridge), they settle within the coelomic epithelium or move through it into the mesenchymal tissues between mesonephros and the coelomic epithelium. Simultaneously, cells of mesonephric origin – deriving from the mesonephric capsules of the glomeruli and/or the mesonephric tubules – move into the same area to mix with the germ cells. In addition cells of the coelomic epithelium move down into this tissue. Thus, the undifferentiated gonad consists of mesenchymal cells and three other cell types: the germ cells, cells derived from mesonephros, and cells originating in the coelomic epithelium. The ingrowing mesonephric cells form cell cords or cell bodies of varying sizes, which connect the mesonephros or the mesonephric tubules and the gonad. As a result of a continuous ingrowth of mesonephric cells and mitotic activity of all gonadal cells, the gonadal ridge soon becomes macroscopically recognizable as an elongated swelling upon the ventral side of mesonephros (Fig. 1.4). In man the gonadal anlage, the forerunner of the gonad, can be distinguished around the sixth week of fetal life, which is simultaneous with the time when the heart begins to beat. In the mouse the gonads are seen from the tenth day of fetal life.

Fig. 1.4. Gonadal swelling on the ventral side of mesonephros from a 15-day-old rabbit fetus.

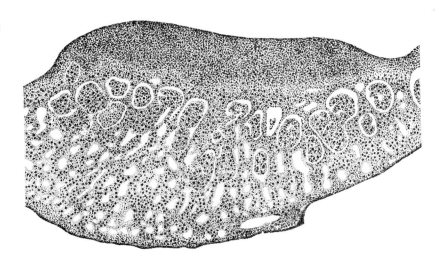

### Differentiation of the gonads

At a certain stage of development the sex of the fetus becomes morphologically distinguishable. The genetic sex is first reflected in the morphology of the gonads. The distribution of the germ cells and the amount of mesonephric tissue growing into the gonad mirrors the early gonadal sex differentiation.

In the undifferentiated gonads, the gonia and the somatic cells are distributed all through the tissue. At the time of sex differentiation, cellular

Fig. 1.5. Early formed testicular cord with large, clear prospermatogonium in a 33-day-old pig fetus. Many Leydig cells (arrowheads) are seen among the solid cords.

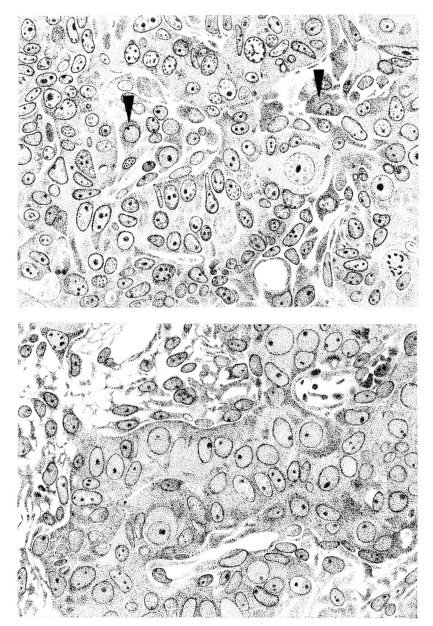

Fig. 1.6. Cell cords with oogonia in an ovary of a 33-day-old pig fetus ('delayed meiosis'). No cells resembling the Leydig cells shown in Fig. 1.5 can be recognized, a fact that makes it possible to determine that this is an ovary.

cords start to develop from the central part of the testis towards the periphery. These testis cords enclose the germ cells, the spermatogonia, together with somatic cells, the Sertoli cells (Fig. 1.5). The testis cords are formed with connections to the ingrowing mesonephric cell cords. These connections remain throughout the whole span of life and constitute the rete testis. Simultaneously with the formation of testicular cords, the ingrowth of mesonephric cells decreases or stops. Further growth of the testis is now mainly dependent on proliferation of the cells already present within the gonad. The morphological differentiation of the testis is closely accompanied by a rapidly increasing capacity for steroid synthesis, particularly testosterone.

The differentiation of the female gonad, including the ability to synthesize steroids, is species dependent. In general there are two types of early ovarian differentiation. In one the germ cells are enclosed in cell cords resembling testicular cords, which are also connected to the mesonephros (Fig. 1.6). This type of ovarian differentiation characterizes development in the sheep and the pig. In other species no cord formation occurs and

Fig. 1.7. Compact ovary with germ cells in different stages of meiosis of a 15-week-old human fetus, showing an example of 'immediate meiosis'.

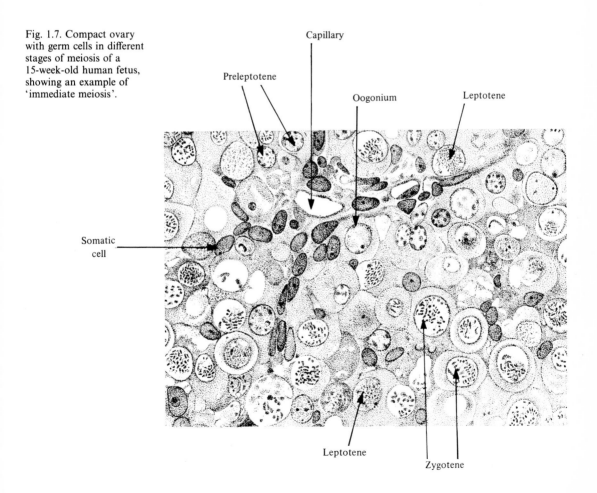

Fig. 1.8. This schematic drawing shows the pattern of differentiation in male and female. The gonadal anlage is formed in close association with the mesonephric tissue. During gonadal growth the germ cells (white spheres) divide mitotically. Simultaneously, cells (indicated by heavy stipple) stream in from the coelomic epithelium and from the mesonephric glomeruli or the mesonephric tubules and surround the germ cells. In the females of some species, such as the sheep and pig, the germ cells become enclosed in cell cords at the time of gonadal sex differentiation, although the start of meiosis occurs somewhat later ('delayed meiosis') (column B). In females of other species, such as mouse and man, the germ cells are not confined to cords, and the ovary is compact, with germ cells uniformly distributed throughout its substance. In such ovaries meiosis starts almost at the same time as gonadal sex differentiation ('immediate meiosis') (column C). In all females the cords disappear when meiosis starts. In males, meiosis is delayed until puberty (column A).

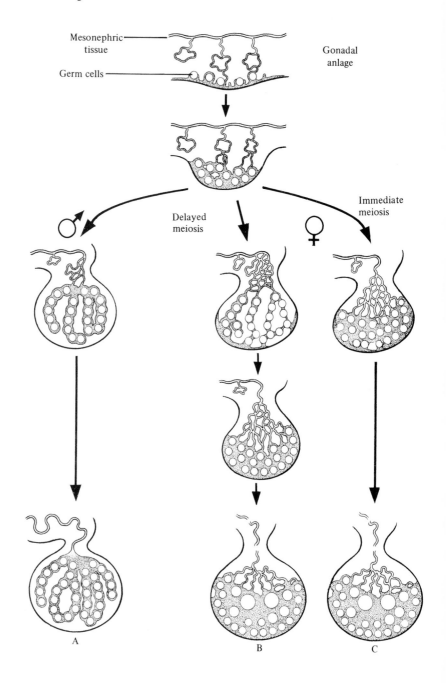

the ovary is a compact organ – a type of development that is seen in man and mouse (Fig. 1.7). In both cases the gonadal–mesonephric connection remains during early ovarian growth.

In contrast to conditions in the testis, mesonephric cells continue for some time to move into the growing ovary. A central basal area of the ovary becomes stuffed with a network of mesonephric cell cords. The centrally situated germ cells then degenerate leaving a sterile central medulla and a germ-cell-rich peripheral cortex. The different patterns of gonadal differentiation are shown schematically in Fig. 1.8.

While the gonads are differentiating, the urogenital system also proceeds through a series of changes from the indifferent condition to the recognizably male or female forms of the organs (Fig. 1.9). The process involves a progressive modification of mesonephric Bowman's capsules; these structures begin in a form very similar to that of the primordia of the Bowman's

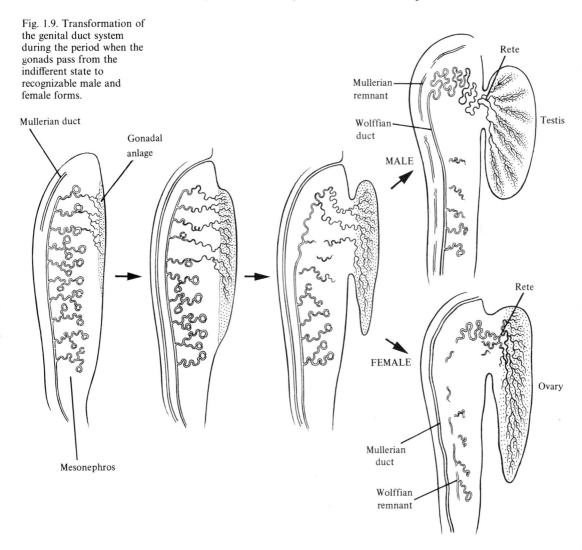

Fig. 1.9. Transformation of the genital duct system during the period when the gonads pass from the indifferent state to recognizable male and female forms.

capsules of the primitive kidney but become transformed into simple tubular structures that give rise to the rete testis and rete ovarii (Fig. 1.10). The connecting tubes are retained only in the male where they become the efferent ducts leading from testis to epididymis.

Although most people agree that the genetic sex chromosome constitution determines the sexual differentiation of the gonads, they still dispute what is triggering this process. A recent proposal is that a specific

Fig. 1.10. Illustrating the way in which the Bowman's capsule of the mesonephros is converted into a simple tube connecting the rete system to the Wolffian duct. In the male these tubes become the efferent ducts and in the female they disappear.

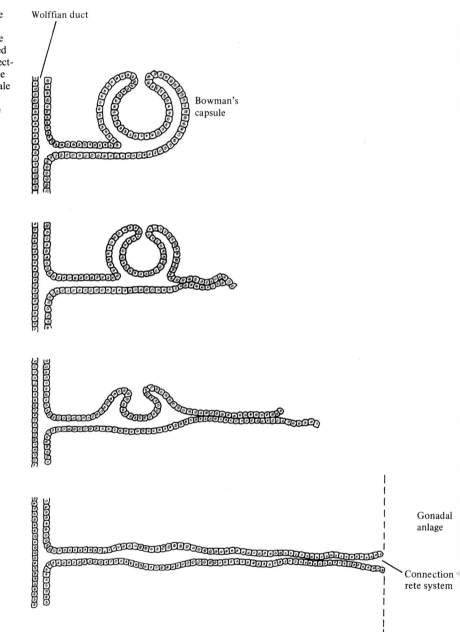

testis-organizing substance, identified as the H-Y antigen, is present on the surface of undifferentiated male gonadal cells, and that this antigen causes the cells to aggregate and form a testis (see Book 2, Chapter 3).

**Regulation of meiosis in males**

In the male the reduction divisions of the germ cells leading to the formation of spermatozoa are not normally initiated until puberty. However, already in fetal or neonatal life the male germ cells have the ability to enter early stages of meiosis. They seem to become capable of entering meiosis at the same time that germ cells of the ovary start meiosis. The fetal male germ cells, which are generally enclosed in testicular cords, only reach the preleptotene or occasionally the leptotene stage, but if, accidentally as it were, some germ cells have not been confined to the cords, but left outside in the surrounding mesenchymal tissue, they may advance into the meiotic prophase and reach zygotene and pachytene stages, especially if they are situated close to the mesonephros (Fig. 1.11). There they degenerate and disappear.

The stimulus to enter meiosis can be ascribed to the action of a postulated diffusable meiosis-inducing substance (MIS), which is secreted by cells derived from the mesonephros. The reason why meiosis does not proceed in the cord-enclosed germ cells until puberty is explained by the action of a postulated meiosis-preventing substance (MPS) present within the cords. The theory is that MIS and MPS oppose each other and that advance into meiosis is dependent on the relative concentrations of the two substances.

That MIS and MPS are secreted by testes of different mammals can be shown in experiments on mouse gonads *in vitro*. Such studies have made it clear that puberal and adult testes secrete MIS, which can induce meiosis in fetal testes. Advanced stages of the meiotic prophase appear in fetal mouse testes when these are grown in media previously used for the culture of puberal and adult testes. The presence of MPS can also be demonstrated *in vitro*: culture media in which fetal testes have grown contain a substance that prevents or inhibits the onset of meiosis in fetal mouse ovaries. The lack of meiosis in fetal and infant testes can therefore reasonably be ascribed to the action of this meiosis-preventing substance.

**Regulation of meiosis in females**

The stage of development at which meiosis starts in the female is species-dependent, although it is always initiated early in life, often before birth. In the ovaries of many species meiosis is initiated simultaneously with or shortly after gonadal sex differentiation occurs, an event that has been termed 'immediate meiosis' (Fig. 1.8). In other species, a 'delaying period' separates gonadal sex differentation and the onset of meiosis. This has been called 'delayed meiosis'. Species with immediate meiosis, such as mouse, guinea pig and man, have compact ovaries with their germ cells

uniformly distributed all over the organ or within a broad cortex. In species
with delayed meiosis, as in the pig, cow, rabbit and mink, the germ cells
are enclosed in more or less well-defined cords during the delaying period.
Species with delayed meiosis synthesize considerable amounts of steroid
(oestradiol) before meiosis starts, whereas in those with immediate meiosis
little or no steroid can be detected before meiosis begins. In all species

Fig. 1.11. Meiosis occurs at
different times and for
different periods in the lives
of female and male germ
cells; the open circles
represent germ cells in
meiosis and the filled
circles germ cells still
undergoing mitotic
division.

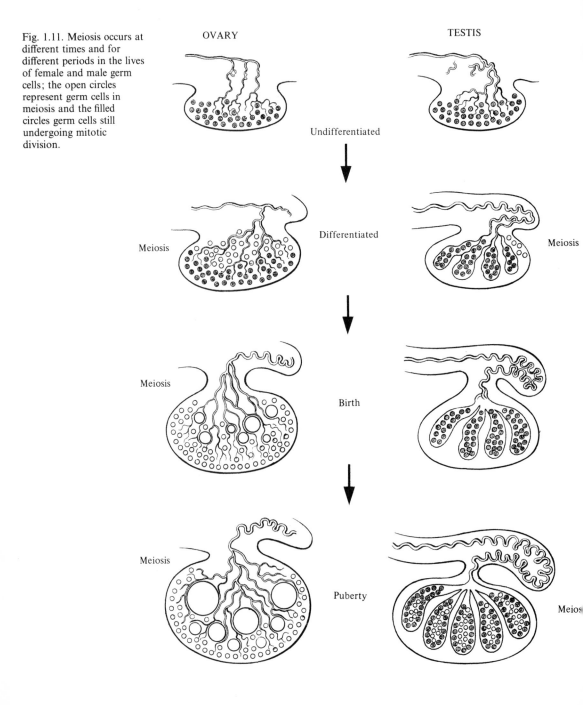

studied, the onset of meiosis is correlated with a low level of steroids within the ovary.

In different species it has been shown that ovaries or their mesonephric systems secrete MIS at the time when meiosis starts. The chemical nature of MIS has only been partially determined, but the suggestion is that its action is closely related to steroid synthesis by the gonad. The reason why meiosis is arrested in prophase is still not understood. Some experiments seem to indicate that the follicle cells exert a meiosis-inhibiting influence on the oocyte.

The DNA of an oocyte is synthesized prior to the first meiotic prophase. The nature of the life cycle of the female germ cell means, therefore, that these cells can carry some very old DNA by the time that they finally come to ovulation and fertilization. In larger mammals this period may amount to many years – ample time for the oocytes to suffer damage from drugs or radiation, the effects being expressed as genetic disturbances in the offspring.

The continuity of germ plasm from generation to generation is one of the most important principles in biology, yet it apparently lacks support in mammalian development. Germ plasm is readily identifiable in the oocyte of amphibians, but there is little evidence for it in mammals. For example, we know that the cells of a $4\frac{1}{2}$-day-old mouse blastocyst are still capable of giving rise to either germinal or somatic elements. So when, and how, do sex and soma first separate? The germ cells are after all the very quintessence of life, and a killjoy could regard soma as merely the vehicle for the transmission of sex from generation to generation. It is tantalizing to realize that we know so little about how mammalian germ cells differentiate from somatic cells in the first place, and why it is that meiosis is their unique prerogative.

## Suggested further reading

Mechanisms of genetic sex determination, gonadal sex differentiation, and germ-cell development in animals. J. R. McCarrey and K. K. Abbott. *Advances in Genetics* **20**, 217–90 (1979).

Migration of the germ cells of human embryos from the yolk sac to the primitive gonadal folds. E. Witschi. *Contributions to Embryology* **32**, 67–80 (1948).

Regulation of meiosis in mammals. A. G. Byskov. *Annales de Biologie Animale, Biochimie et Biophysique* **19**, 1251–61 (1979).

Continuity of the female germ cell line from embryo to adult. B. Mintz. *Archives d'Anatomie Microscopique et Morphologie Experimentale* **48**, 155–72 (1959).

Origin and migration of primordial germ cells in mammals. E. M. Eddy, J. M. Clark, D. Gong and B. A. Fenderson. *Gamete Research* **4**, 333–62 (1981).

Primordial germ cells and the vertebrate germ line. M. W. Hardisty. In *The*

*Vertebrate Ovary*, pp. 1–45. Ed. R. E. Jones, Plenum Press; New York and London (1978).

Oogenesis and ovarian development. T. G. Baker. In *Reproductive Biology*, pp. 398–437. Ed. H. Balin and S. Glasser. Excerpta Medica; Amsterdam (1972).

*Reproduction*. J. Cohen. Butterworths; London and Boston (1977) (especially chapters 1 and 4).

# 2

# Oogenesis and ovulation

*T. G. BAKER*

Oogenesis is the story of the formation, growth and maturing of the female gamete. As we saw in Chapter 1, the process begins in embryonic life, continues after birth (accelerating during puberty) and reaches a climax at the time of ovulation. By far the most important change undergone is meiosis, which is in a sense the antithesis of fertilization: it halves the number of chromosomes while fertilization restores the diploid complement.

Ovulation releases the egg from the ovary at the right time for its fertilization and further development. Ovulation requires the rupture of the Graafian follicle, which has developed under the control of gonado-

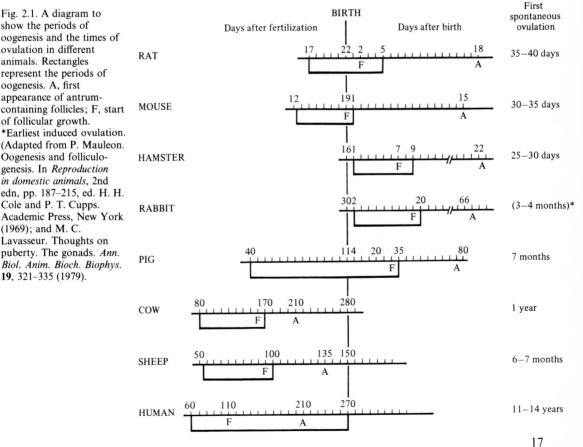

Fig. 2.1. A diagram to show the periods of oogenesis and the times of ovulation in different animals. Rectangles represent the periods of oogenesis. A, first appearance of antrum-containing follicles; F, start of follicular growth. *Earliest induced ovulation. (Adapted from P. Mauleon. Oogenesis and folliculo-genesis. In *Reproduction in domestic animals*, 2nd edn, pp. 187–215, ed. H. H. Cole and P. T. Cupps. Academic Press, New York (1969); and M. C. Lavasseur. Thoughts on puberty. The gonads. *Ann. Biol. Anim. Bioch. Biophys.* **19**, 321–335 (1979).

17

trophic hormones from the pituitary gland and depends on a particular endocrine balance that was necessary for full follicular growth to occur. As we shall see later, the final stages of meiosis need the completion of follicular growth and the release of pituitary hormones. The time relations of oogenesis and ovulation vary a good deal in different species (Fig. 2.1).

**Anatomy of the female reproductive organs**
The positions and relationships of the generative organs in a woman are depicted in Figs. 2.2 and 2.3. The ovaries are attached to the posterior body wall within the pelvic cavity. Close by are the fimbriae of the Fallopian tubes (or oviducts) which communicate with the body of the uterus (a single structure in primates; two uterine horns are found in most other mammals). The Fallopian tubes are the sites of fertilization (within the ampullae) and also serve to regulate the time taken for the fertilized embryo to reach the uterus (about five days in man). There will be more information on these things in later chapters. The dorsal wall of the uterine body is the usual site of implantation, and at parturition the mature fetus

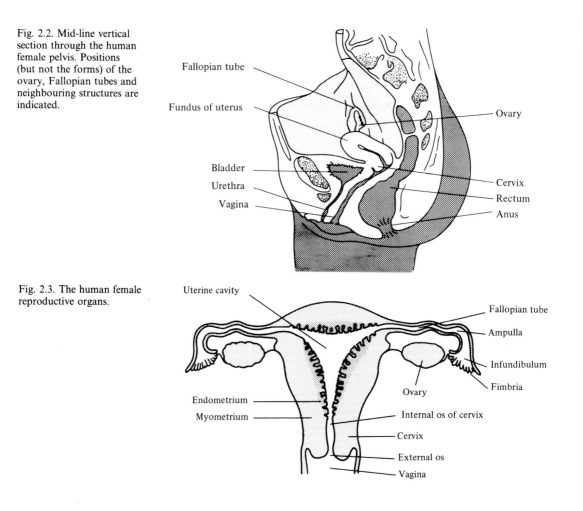

Fig. 2.2. Mid-line vertical section through the human female pelvis. Positions (but not the forms) of the ovary, Fallopian tubes and neighbouring structures are indicated.

Fallopian tube

Fundus of uterus

Ovary

Bladder

Urethra

Vagina

Cervix

Rectum

Anus

Fig. 2.3. The human female reproductive organs.

Uterine cavity

Fallopian tube

Ampulla

Infundibulum

Fimbria

Endometrium

Myometrium

Ovary

Internal os of cervix

Cervix

External os

Vagina

is expelled through the cervix and vagina. The cervix contains mucus which may serve as a sperm reservoir, while the vagina is the copulatory organ of the female and terminates at the exterior with folds of skin (the labia minora and majora).

### Structure of the ovary

Fig. 2.4 is a diagram of a mammalian ovary, showing stages in the production of follicles and the corpus luteum. The follicles contain the oocytes and serve as the vehicles for ovulation as well as producing steroid hormones (see Book 3). After ovulation the remains of the follicle form the corpus luteum which itself produces steroid hormones (largely progesterone).

Chapter 1 was concerned with the origin of the germ cells and the process

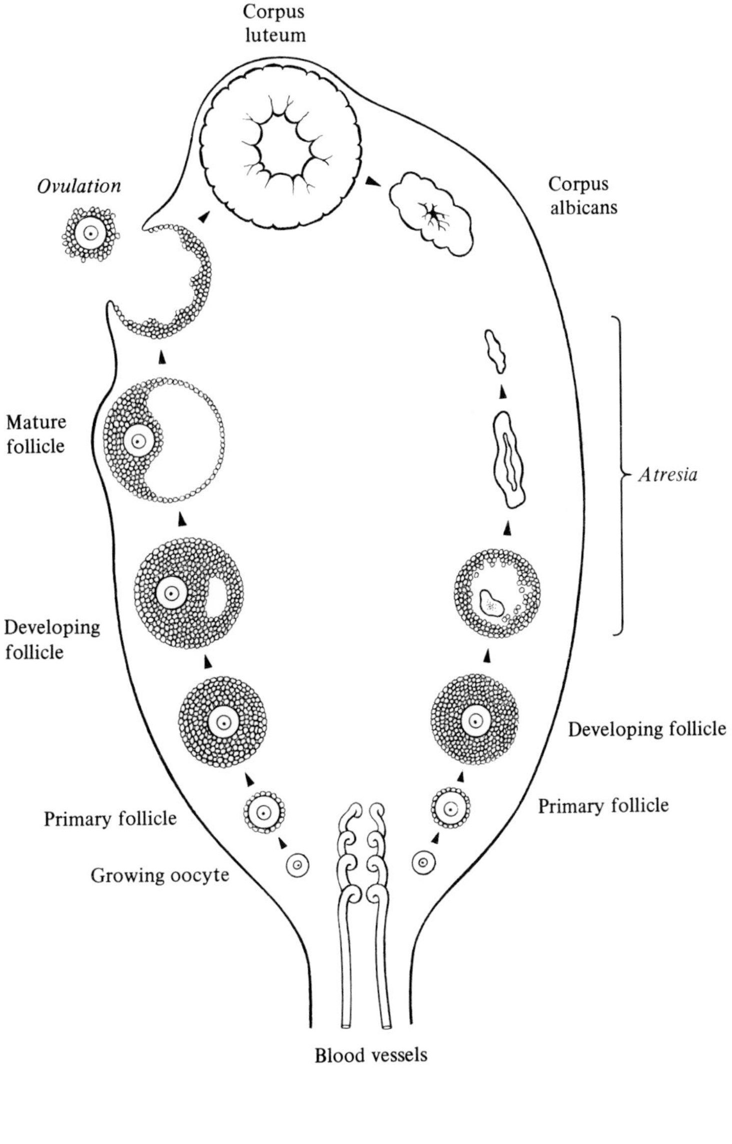

Fig. 2.4. Diagram of a mammalian ovary showing, on the left, the development of a primordial follicle to the stage of ovulation, after which a corpus luteum forms and ultimately regresses. On the right is shown the much more common occurrence of development terminated by atresia. (From R. Meadows. *Pocket Atlas of Human Histology.* Oxford University Press (1980).)

of sexual differentiation of the gonads; in the ensuing sections of the present chapter we will consider the steps in the development and maturation of the germ cells and their follicles, the mechanism of ovulation, the development of the corpus luteum and the occurrence of germ cell degeneration (atresia).

### Life history of female germ cells

In Chapter 1 the point was made that the earliest recognizable germinal 'stem' cell is the primordial germ cell which is sometimes called a gonocyte. Depending on the genotype of the individual, and on certain genes that can modify sexual differentiation (see Book 2, Chapter 3), these cells differentiate into either oogonia in a female or spermatogonia in a male.

Fig. 2.5. Life-cycle of the female germ cell.

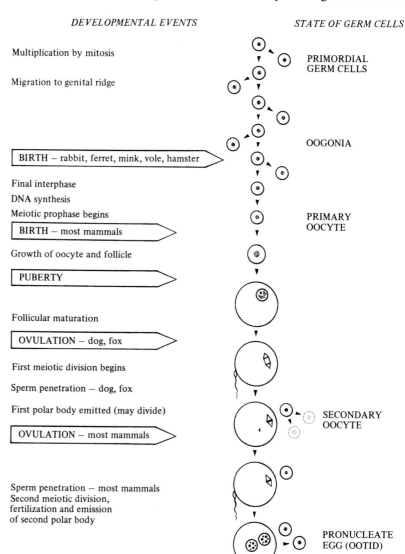

*DEVELOPMENTAL EVENTS*  *STATE OF GERM CELLS*

Multiplication by mitosis

PRIMORDIAL GERM CELLS

Migration to genital ridge

OOGONIA

BIRTH – rabbit, ferret, mink, vole, hamster

Final interphase

DNA synthesis

Meiotic prophase begins

PRIMARY OOCYTE

BIRTH – most mammals

Growth of oocyte and follicle

PUBERTY

Follicular maturation

OVULATION – dog, fox

First meiotic division begins

Sperm penetration – dog, fox

First polar body emitted (may divide)

SECONDARY OOCYTE

OVULATION – most mammals

Sperm penetration – most mammals
Second meiotic division,
fertilization and emission
of second polar body

PRONUCLEATE EGG (OOTID)

All of these cell types are capable of mitosis, although the actual number of cell divisions (particularly for oogonia) may be finite. Once the oogonium has completed its last pre-meiotic division it becomes an oocyte and enters meiosis (see below), firstly as a primary oocyte (first meiotic division) and subsequently as a secondary oocyte (second meiotic division) (see Fig. 2.5).

*Meiotic changes in oocytes*

Meiosis consists of two cell divisions, the first of which involves the halving of the chromosomal number (diploid to haploid) and permits the exchange of genetic information between pairs of chromosomes, one member of each pair having been derived from the mother and the other from the father. The second division resembles mitosis, although only haploid chromosomes are involved. In the human female, all the cells of the body (including oogonia) undergo mitotic divisions, the number of chromosomes remaining constant at 46. Primary oocytes are germ cells in which the first meiotic division occurs, resulting in secondary oocytes with only 23 chromosomes. Each chromosome is split longitudinally into a pair of chromatids which separate during the second meiotic division, so that the egg is left with single (unpaired) chromosomal threads (see Fig. 2.5). Both of these nuclear divisions are accompanied by unequal cytoplasmic divisions, so that two small polar bodies are formed.

Cell division – whether mitotic or meiotic – proceeds in turn through prophase, metaphase, anaphase and telophase (Fig. 2.6). The terms used for subdividing cell division are purely arbitrary, however, since the process is dynamic and proceeds with few interruptions once started. The interval between successive cell cycles is termed interphase, and is important since it is then that DNA replication occurs (see Chapter 1). The prophase of the first meiotic division is exceptional in that it is greatly prolonged, is highly specialized, and in the female is interrupted by two periods of arrested development (the dictyate stage and either metaphase I or II, to which we will return later). This prophase can be conveniently divided into five stages identified by cytologists as leptotene, zygotene, pachytene, diplotene and diakinesis (the last being the so-called 'germinal vesicle' stage).

Shortly after the last mitotic division by the oogonium (see Chapter 1), the cell enters an interphase in which DNA is replicated in preparation for meiosis. This is the only time in the first meiotic division that the germ cells incorporate such radioactive precursors of DNA as $^3$H-labelled thymidine. During this stage (which is often called pre-leptotene) the nucleus acquires a speckled appearance and short segments of chromosomal threads may be seen. The cell enlarges and shows an increased affinity for nuclear dyes until eventually the diploid number of chromosomes can be identified in cytological preparations (leptotene stage; see Fig. 2.6). Let us consider a hypothetical example in which the germ cells contain four chromosomes, as shown in the illustration. There are two long chromosomes and two short chromosomes; one of each type is derived from the mother, the others from the father. These are the homologous maternal and paternal chromosomes, which during the zygotene stage of meiosis are attached together to form pairs. The pairing process (or synapsis as it was called in the older literature) may begin either at the ends of the chromosomes

Fig. 2.6. Diagram showing the behaviour of chromosomes during meiosis. For simplicity only two pairs of chromosomes are depicted, a short pair with terminal pairing, and a long pair with intermediate synapsis. Only the nuclei are shown; the cytoplasm of the cells has been disregarded. 1, leptotene; 2, zygotene; 3, pachytene; 4, late pachytene, showing tetrads (two pairs of bivalents); 5, diplotene; 6, diakinesis; 7, metaphase I; 8, late anaphase I; 9, telophase I (egg and polar body); 10, metaphase II; 11, anaphase II; 12, telophase II (egg and polar body).

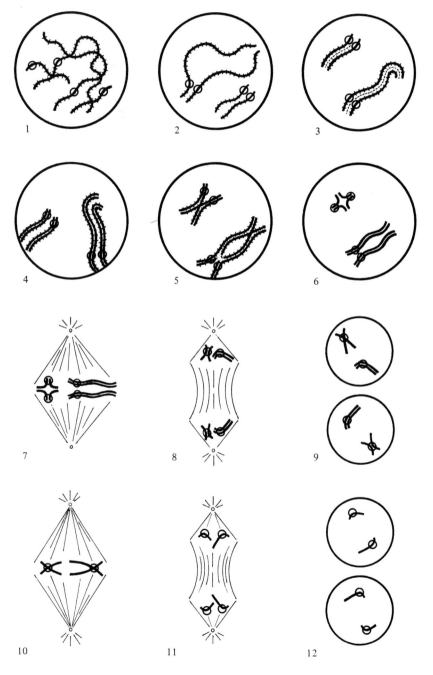

(terminal pairing), or at some point in between (intermediate). Pairing is so precise that homologous genes become associated with one another; if a small region of one chromosome (say the maternal partner) is missing, the paternal thread will form a loop to maintain the correct gene sequence (see Fig. 2.7) but this is not a feature of normal meiosis. Pairing is accomplished in a short period of time and hence the zygotene stage is transitory. In contrast, pachytene, which rapidly ensues, is of relatively long duration. The chromosomes are now paired along their entire length and they shorten and thicken by spiralization of the threads (Fig. 2.6). Studies with the electron microscope have shown that the 'synaptonemal complexes' thus formed each consist of two outer threads of chromosomal material separated by a fine association line.

As the cell grows and gains a great complexity of cytoplasmic organelles (seemingly the 'late' pachytene stage), the lateral arms of the synaptonemal complex can be seen to be split longitudinally to form four threads arranged in pairs (the complexes now being termed tetrads or bivalents), while the intermediate line remains unchanged. These threads break and rejoin during spiralization, and parts of the maternal and paternal homologues are exchanged. The changes are difficult to detect in cytological preparations and are inferred in retrospect from studies of oocytes at the subsequent stage, diplotene. The chromosomes now show mutual repulsion from each other so that the bivalents separate except at certain points (chiasmata; Figs. 2.6 and 2.8) where 'crossing-over' has occurred during the phase of chromosomal breakage and reunion. The exchange of genetic material between maternal and paternal threads results in a reassortment of genes, which ensures that the chromosomes of the oocyte are different from those of either parent. The homologous chromatids constituting each bivalent still remain attracted to each other, and hence the pattern characteristic of the diplotene stage is maintained (see Fig. 2.8).

This description of meiotic prophase so far applies equally well to oogenesis and spermatogenesis, but now the similarities end. Meiotic

Fig. 2.7. Diagram showing precision of pairing at zygotene. Individual genes on maternal and paternal chromosomes are paired, apart from '*d–h*' which are missing from one homologue.

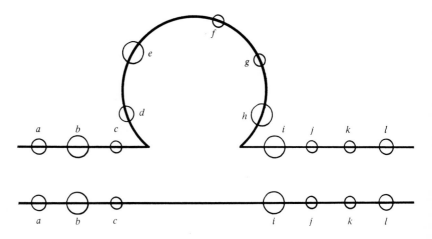

prophase in the female is completed to the diplotene stage shortly after birth in most species, and the cell enlarges greatly. Oocytes then enter a prolonged 'resting phase' which is terminated shortly before ovulation with pre-ovulatory meiotic changes in the Graafian follicle. In such species as man, the earliest viable oocyte to resume meiosis does so at the time of puberty, and the last egg to pass through pre-ovulatory maturation may be found in women aged 45–50 years. That this long period of arrested development may be of great importance is suggested by the observation that the incidence of embryos with meiotic defects (e.g. the extra chromosome number 21 in mongolism) increases with maternal age, but there are other possible explanations for these defects (see Book 2, Chapter 5).

By contrast, meiosis in the male does not begin until around the time of puberty, but then continues without interruption throughout adult life. There is no period of arrested development in the spermatocyte, diakinesis following immediately after the diplotene stage (see Chapter 4).

These differences have a profound effect on the number of germ cells available for reproduction. In the male, the number of spermatozoa that can be produced is vast since the stem cells, or spermatogonia, are being continually replaced throughout life by mitosis. The stem cells of the ovary (oogonia) have a limited life and probably a finite number of mitotic divisions. Consequently, the population of germ cells in the female has a fixed upper limit which, with the disappearance of oogonia, is rapidly depleted with increasing age by the process of atresia. By way of example we may refer again to the situation in the human female (see also Chapter 1): the number of germ cells increases from about 1700 during migration to some 600 000 during the second month of gestation, and subsequently to almost 7 000 000 by mid-term (see Fig. 2.9). The population of germ cells then declines rapidly to about two million (of which half are already degenerate) at the time of birth. This decline, as we shall see later in this chapter (p. 42), is due mainly to the elimination of large numbers of cells by the process of atresia. But a contributing factor is the eventual cessation of mitosis in oogonia and their transformation into oocytes at the leptotene stage of meiosis.

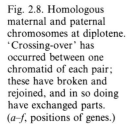

Fig. 2.8. Homologous maternal and paternal chromosomes at diplotene. 'Crossing-over' has occurred between one chromatid of each pair; these have broken and rejoined, and in so doing have exchanged parts. (*a–f*, positions of genes.)

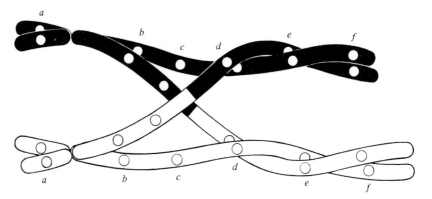

The number of oocytes continues to decline with increasing age until the time of the menopause in women (or the cessation of reproductive function in animals), when few oocytes can be detected in histological sections of the ovary. Of the seven million oocytes that were at one time present in the human ovary only about 400 to 500 will have been ovulated. The remainder of the germ cells, like the vast majority of spermatozoa in the male, will have fallen by the wayside in what must appear to be a very wasteful process in reproduction. This conclusion becomes firmer when one considers that the number of children born to a couple could hardly exceed thirty. As we shall see in later chapters, the process is not quite so wasteful as these figures suggest since numerous other factors are involved that limit both the number of viable offspring and the timing of successful conception and implantation.

*The dictyate, diffuse diplotene, or dictyotene stage of meiotic prophase*
In the rat the process of meiosis in the ovary is synchronized and 90 per cent of the germ cells at any particular time are at the same stage of development. In other species, asynchrony is evident and the ovaries of fetuses contain germ cells at all stages of mitosis and meiosis.

By the time of birth, the ovaries of rats, guinea pigs, sheep, cows, monkeys and human beings contain mainly oocytes that have reached the

Fig. 2.9. Changes in the total population of germ cells in the human ovary with increasing age. (From T. G. Baker, *Am. J. Obstet. Gynec.* **110**, 746 (1971). Data for children and adults from E. Block, *Acta Anat.* **14**, 108 (1952).)

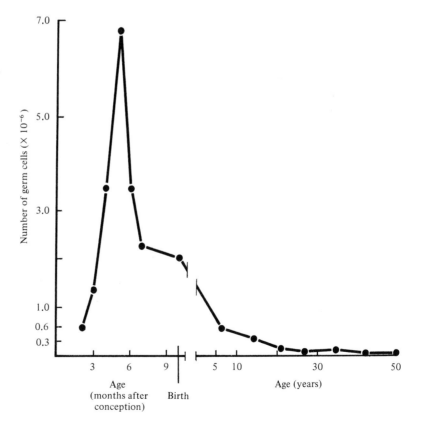

diplotene stage of meiosis. In contrast, those of the newly born rabbit, ferret, mink, vole and golden hamster contain only oogonia, the prophase of meiosis being completed within the first few weeks after birth. Oogenesis in the pig occurs mainly in fetal life but extends into the post-natal period, while in the cat oocytes at early stages of meiotic prophase may be found up to the time of puberty. Inbreeding can affect the temporal relations of oocytes: CBA and A strains of laboratory mice contain oocytes that have reached the diplotene stage at birth, while those of the Street and Bagg strains are mainly at pachytene.

In all these species, irrespective of the timing of oogenesis, oocytes embark on a prolonged 'resting' phase shortly after the onset of diplotene. This so-called dictyate or dictyotene stage is characterized by highly diffuse chromosomes, the DNA of which has little affinity for such nuclear strains as Feulgen's reagent. In the extreme condition, found in oocytes of rats and mice, the nucleus contains what seems superficially to be a reticulum of fine threads. Some authors believed that these threads 'disappeared', to be reconstituted at a later date from within the nuclear sap, and also (largely on the basis of inadequate or imprecise histochemical techniques) that the oocyte contained virtually no RNA and thus lacked the ability to manufacture the protein and other materials that can loosely be grouped as 'yolky substances'. For these reasons the period of arrested development was considered to be truly a resting phase in which the oocyte was merely 'nursed' by its investing follicle cells, its growth being only passive.

There can be no doubt that oocytes at the dictyate stage grow considerably, before as well as after the onset of follicular growth. But follicle cells are not necessary for the growth of the oocyte since their absence is occasionally observed following irradiation or hypophysectomy. Moreover 'giant oocytes', whose granulosa envelope remains undeveloped, have been reported for numerous species including mouse, human and rhesus monkey.

Studies with the electron microscope have shown that both the oocyte *and* its granulosa cells contain the cytoplasmic organelles that are usually associated with cells secreting proteins and mucopolysaccharides. Thus oocytes contain Golgi vesicles, endoplasmic reticulum (mainly of the smooth type), and abundant rosettes of ribosomes, as well as occasional vesicles of what could conceivably be the products or precursors of secretion. Unfortunately, the use of radioactive tracer substances has so far proved of little value in assessing the roles of the oocyte and granulosa cells in the growth phase, since the precursors involved are incorporated by almost all cells within the ovary, including oocytes.

Studies on oocytes at the dictyate stage have shown that the chromosomes bear lateral projections in the form of branches and loops (Fig. 2.10) which actively replicate ribonucleic acid (RNA). They closely resemble the 'lampbrush' chromosomes, which are found almost universally in eggs of lower vertebrates and some invertebrates. The RNA may act as the messenger directing protein synthesis within the oocyte itself. Circumstantial evidence from studies of oocytes in frogs and toads suggests that some of the RNA may act as the early 'organizer' of mammalian development. Whatever the outcome of the additional studies required to elucidate these complex problems, the dictyate stage should clearly not be referred to as a resting phase, since the oocytes show a high degree of metabolic and synthetic activity at a time when the follicular envelope consists of only a few flattened epithelial cells. Only during the second half of the oocyte's growth phase do the granulosa cells clearly contribute maternal protein and other materials to the ooplasm (see later).

*Formation and function of the zona pellucida*

The zona pellucida (Fig. 2.11) consists of mucopolysaccharide and trypsin-digestible material, and is indistinct in sectioned material unless stained by specific histochemical stains (as in the periodic acid–Schiff (PAS) reaction). When viewed with the electron microscope it appears somewhat 'fluffy' due to precipitation of its protein component with the fixatives employed.

The zona pellucida forms around oocytes that are surrounded by a complete layer of cuboidal granulosa cells (Fig. 2.12). Islands of fibrillar

Fig. 2.10. Diagrammatic representation of a chromosome at the diplotene/dictyate stage of meiotic prophase. Fibrils (B) constituting loops and lateral projections arise from the bivalent (A). Granular material containing ribonucleo-protein is associated with the ends of the projections (C) and loops (D). (From T. G. Baker and L. L. Franchi. *Chromosoma* **22**, 358 (1967).)

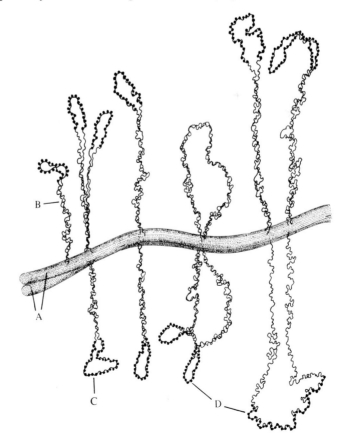

material are deposited in spaces between adjacent granulosa cells and the oocyte surface. The source of the fibrils is obscure, although islands of material within the endoplasmic reticulum of both the follicle cells and the oocyte may represent precursor substances. The cellular transformation of proteinaceous substances into mucopolysaccharides is known to occur in Golgi vesicles, which are usually found close to the oocyte surface in regions where the zona is being deposited. Furthermore, the material of the zona pellucida is not homogeneous but consists of two layers, the outer staining more intensely than the inner, and so the zona pellucida may well be the product of both the oocyte and its follicular envelope. Autoradiographic studies with [35]S-labelled amino acids have confirmed the role of granulosa cells in the processes involved, but further studies are required to determine the part played by the oocyte.

The islands of zona material fuse together and eventually form an impervious jelly-like coat which completely surrounds the egg. The structure in some ways resembles a sieve, in that microvilli from the oocyte surface and coarser processes from the granulosa cells occupy canals within

Fig. 2.11. Structure of fully formed zona pellucida (ZP) around an oocyte in a Graafian follicle. It is penetrated by microvilli (M) arising from the oocyte on one side and by processes from the granulosa cells (G) on the other. These processes indent the cytoplasm of the oocyte and may provide nutrients and maternal protein. N, oocyte nucleus.

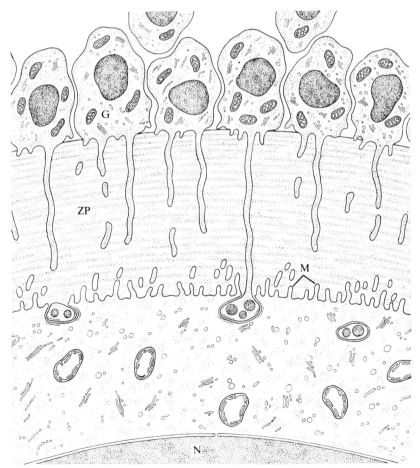

the zona (see Chapter 3). The granulosa-cell processes often traverse some distance into the oocyte and may transfer maternal protein into vesicles that are budded off into the cytoplasm. The granulosa cells, which later become transformed into the cumulus oophorus, are clearly essential for the nutrition of the egg once the zona pellucida is established; they produce metabolites such as lactic and pyruvic acids.

## Follicular growth

Soon after the oocytes are formed they become surrounded by a single layer of flattened epithelial cells which arise either from the rete ovarii (see Chapter 1) or from the coelomic covering of the ovary. Follicular growth involves a change in the shape of the epithelial cells, which become cuboidal in form, and an increase in their number by mitosis. The follicle thus becomes two- and then three-layered, and after a finite number of divisions, fluid accumulates in spaces between the epithelial cells and gathers to form a single antrum; the follicle is now described as being vesicular (Fig. 2.13).

The results of recent experiments have shown that a short-lived surge of gonadotrophic hormones is required to initiate the process of follicular growth. If this surge is blocked in mice by the injection of antisera to gonadotrophins shortly after birth, or if ovaries maintained in culture are deprived of the hormones, follicles fail to form more than a single layer of granulosa cells. In the human fetus lacking a hypothalamus (anencephalic

Fig. 2.12. Drawing from an electron micrograph showing the initial deposition of zona material (Z) between two granulosa cells in a very early follicle. B, basement membrane; O, oocyte.

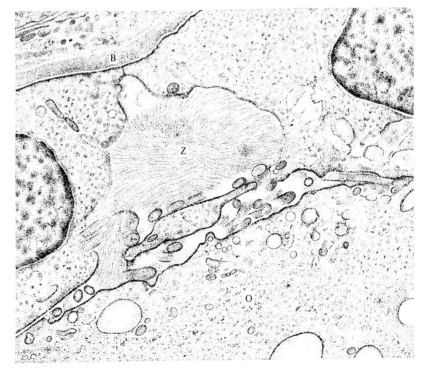

fetus) 'growing' and antral follicles fail to develop, whereas normal fetuses often have many vesicular follicles by the time of birth. It should be pointed out, however, that once this initial 'triggering' of follicular growth has been accomplished the process continues throughout reproductive life, and there is no evidence that gonadotrophic hormones are required to sustain follicles with less than four layers of granulosa cells (see below).

The ovary is a dynamic structure in which vesicular follicles are constantly developing from those of the primordial type (Fig. 2.13). The first 'growing' follicles appear in the ovaries within a few days of birth in most species: in primates they are found before birth, together with small vesicular follicles. But only with the establishment of the correct hormonal balance, and the initiation of reproductive cycles at the time of puberty, is

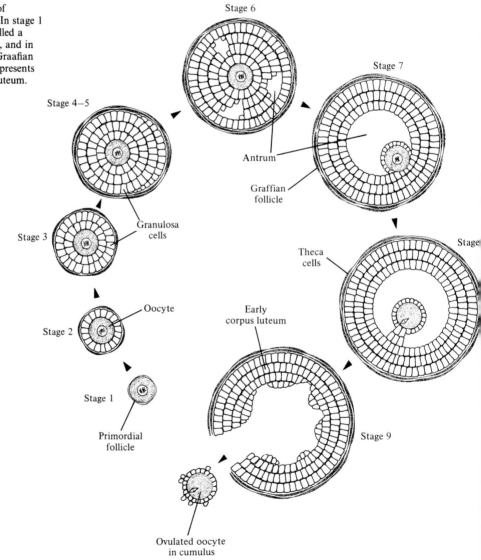

Fig. 2.13. Stages of follicular growth. In stage 1 the structure is called a primordial follicle, and in stages 7 and 8 a Graafian follicle. Stage 9 represents the early corpus luteum.

Stage 6

Stage 7

Stage 4–5

Antrum

Graffian follicle

Stage 3

Granulosa cells

Stage

Theca cells

Oocyte

Early corpus luteum

Stage 2

Stage 1

Primordial follicle

Stage 9

Ovulated oocyte in cumulus

the process of follicular growth permitted to culminate in ovulation. (The hormonal control of follicular growth and ovulation is dealt with more fully in Book 3.) The great majority of vesicular follicles that are produced after puberty (and all of those before puberty) undergo degeneration at varying stages in their formation (see p. 43). The number of follicles that attain ovulation is more or less fixed for the species by the levels of circulating gonadotrophic hormones, a point that we shall return to later. Thus, in the human female only one follicle usually undergoes ovulation each month; the remaining twenty or so that have reached the same stage of growth degenerate. The injection of additional hormones causes more follicles to ovulate, a technique which has applications in agriculture and has inadvertently resulted in undesirable multiple births in women treated with gonadotrophins.

About 90 per cent of all oocytes in the ovaries of sexually mature mice, rats and monkeys are enclosed within primordial follicles consisting of one layer of flattened granulosa (epithelial) cells. This follicular envelope is initially incomplete, the cells being scattered about the periphery of the oocyte. The remaining 10 per cent are the 'growing' and Graafian follicles which are often classified for convenience by histological criteria, such as: (i) size; (ii) numbers of granulosa cells (or layers of such cells) within the membrana granulosa; (iii) development of the theca, and (iv) position of the oocyte within its surrounding body of granulosa cells, the cumulus oophorus. The schemes devised for identifying follicular stage vary in their complexity, that shown in Fig. 2.13 being a simplified type that can be used for a wide range of species.

Fig. 2.14. A diagram to illustrate the idea that follicles develop in frequent succession from the pool of non-growing follicles. Most of these become atretic, but if they should reach a critical size at a time when the appropriate balance of gonadotrophic hormones exists in the circulation, they are 'rescued' and can reach full development, followed by ovulation and the formation of a corpus luteum. The interplay of gonadotrophins and oestrogen necessary to induce these events is shown. −ve Oe and +ve Oe, negative and positive feed-back effects of oestrogen. (From H. Peters and K. P. McNatty. (1980) – see Suggested Further Reading.)

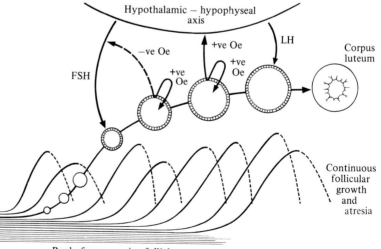

The earliest signs of follicular growth in primordial follicles are: (i) an increase in the size of the oocyte, (ii) a change in shape of the granulosa cells from flat to cuboidal, after which they show frequent mitotic divisions, and (iii) the formation of the zona pellucida. Progressively, the band of granulosa cells, the membrana granulosa, becomes two-, three- and then four-layered. At this time blood capillaries invade the fibrous investment of cells around the follicle and form a vascular coat, the theca interna. This is surrounded in turn by fibroblasts of the theca externa and is the only source of nutrients both for the membrana granulosa and the oocyte.

During each reproductive cycle, as a direct consequence of the release of sufficient FSH by the pituitary gland, a crop of 'growing' follicles (stages 4–5 in Fig. 2.13) are stimulated to undergo further growth and maturation. The number of follicles that are thus 'selected' is determined by the available quantity of gonadotrophin (Fig. 2.14). The pituitary-dependent phase of follicular growth involves further multiplication in the number of granulosa cells, and also the passage of fluid ('liquor folliculi') into spaces between them (stage 6 in Fig. 2.13). This fluid resembles blood serum in composition and is probably derived directly from the capillaries with only slight modification by the follicle cells. As the quantity of fluid increases, the cavities that it occupies increase in size and become confluent to form the antrum. The follicle is now said to be of the Graafian type, after Regnier de Graaf (1641–73) who first described them adequately. With the further expansion of the antrum, the oocyte occupies a position at one side of the follicle and is surrounded by two or more layers of granulosa cells. The innermost layer of these cells becomes columnar in shape and constitutes the corona radiata which, as the innermost part of the cumulus oophorus, persists around the egg for a period after ovulation. The dissolution of these cells whilst the egg is still in the follicle, or precociously after its departure, is a sure sign that degenerative changes are occurring that will result in the death of the oocyte.

Our knowledge of the timing of stages in follicular growth has recently been advanced through studies of the incorporation of [3]H-labelled thymidine into granulosa cells. With each cell division, the number of silver grains over the cell in autoradiographic preparations is reduced by a half. Hence the transit times for each follicular stage can be determined from grain counts and used to estimate fluctuations in the population at each stage of growth. Such studies have shown that follicular growth is a continuous process and that the number of follicles at each of the stages from 1 to 4 fluctuates only slightly with the phases of the reproductive cycle. The populations are thus in a 'steady state', and follicles progress to the subsequent stage only to maintain the number at that stage. But with the onset of stage 5, the numbers of follicles fluctuate widely, being most common during the early part of the cycle and least common after ovulation.

The factors initiating follicular growth in primordial follicles remain completely unknown at present. It is difficult to explain how a few such follicles are 'selected' to embark on the growth phase, while seemingly identical neighbouring follicles remain unaffected. The earliest signs of growth almost certainly occur in the oocyte itself which may trigger the changes in granulosa cells. The oocyte may produce a 'messenger' (possibly RNA from lampbrush chromosomes, as described earlier) at a specific time in its growth phase. Alternatively, the oocytes may be 'programmed' in some way. Perhaps the time of onset of meiotic prophase in germ cells during embryonic or fetal life affects the timing of their subsequent growth phase. The first cells to enter meiosis would also be the earliest to embark on follicular growth. This 'production-line' hypothesis may in part account for germinal selection in the ovaries of cows, monkeys and women in which meiosis is not synchronized and occupies a period of months in the life of the fetus; but it can be of little value in other species where meiotic prophase is completed in synchrony within at most a few days (e.g. mouse and rat). Pituitary hormones are unlikely to initiate follicular growth since the initial phases of the growth process continue after removal of the pituitary gland. Growth up to four layers of granulosa cells is generally held to be independent of hormonal control, while the process beyond this stage is hormone-dependent (see below). Paradoxically, the injection of antibodies to gonadotrophic hormones into mice during the first two weeks of life is said to disturb the normal pattern of development in granulosa and thecal cells, but does not affect growth of the oocyte. Clearly the regulation of the process of follicular growth is highly complex and requires the elaboration of new techniques if the mechanism is to be further resolved.

In contrast, our knowledge of the development of the Graafian follicles from the 'precursor pool' (stages 6–8 in Fig. 2.13) has been considerably enhanced from studies of ovaries in organ culture, and from those involving replacement therapy in animals whose pituitary glands have been surgically removed. In both situations, the injection of follicle-stimulating hormone preparations from pregnant mares' serum and from pituitary glands, and of natural and synthetic oestrogens, can promote further growth of the follicle and antrum formation. Relatively pure preparations of pituitary follicle-stimulating hormone (FSH) are said to be less effective than those containing luteinizing hormone (LH); the optimum concentrations may be about 1:3, respectively. We also know that the action of pregnant mares' serum gonadotrophin (PMSG) on follicular growth in hypophysectomized rats is potentiated by pretreatment with oestrogens. The principal controlling mechanism in the final growth phase of the follicle and its antrum evidently involves a peak of FSH in the presence of some LH, the function of which may be to ensure adequate steroid synthesis (especially of oestrogens). The combined effects of these hormones determines not only the *number* of follicles that develop to the mature Graafian type, but also the *proportion* that undergo ovulation compared with those that degenerate (Fig. 2.14).

*Pre-ovulatory maturation*

If ovulation is to occur, the Graafian follicle (and the egg within it) must undergo further changes that can only take place under precisely controlled hormonal conditions. The level of circulating FSH (Fig. 2.15) is elevated for only a short time at the beginning of the growth phase, after which the quantity of circulating gonadotrophin (FSH and LH) remains fairly constant until a short time before the impending ovulation. The interval between the LH peak and ovulation varies from about 12–15 hours in the mouse, rat and rabbit, through 36 hours in the human female to about 42 hours in the pig. The onset of pre-ovulatory maturation is marked by a sudden and dramatic rise in the release of gonadotrophins from the pituitary, especially of LH (the so-called 'LH surge'). Though both the FSH and LH levels rise (Fig. 2.15), the ratio of these hormones one to the other is probably just as important in controlling pre-ovulatory maturation as the level of LH alone. In any event, these hormones reach an optimum value for each species which affects the final maturation of the oocyte and its surrounding follicles; they also regulate to some extent the number of eggs that ovulate, which is more or less constant for each

Fig. 2.15. Plasma levels of gonadotrophins (solid lines), oestradiol (dashed line) and progesterone (dotted line) during the human menstrual cycle. The cycle is centred on day 0, the day of the mid-cycle peak of LH. (From L. Speroff and R. L. Vande Wiele. *Am. J. Obstet. Gynec.* **109**, 234 (1971).)

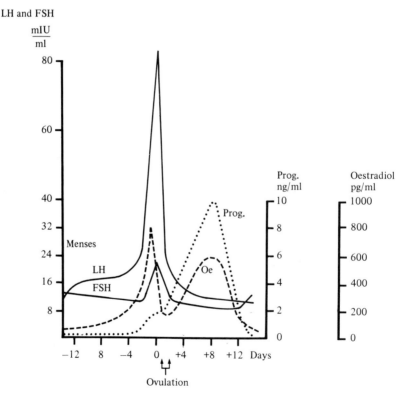

species. Shortly before the gonadotrophin peak, there is a sharp increase in oestrogen level and this is considered to be the stimulus for its production.

The LH surge (or the increased oestrogen that causes it) is said to induce a final 'wave' of mitosis in granulosa cells, so that their number reaches an optimum for the ovulatory Graafian follicle. The quantity of follicular fluid in the antrum also increases dramatically; owing to increased permeability of the blood–follicle barrier, the intra-follicular pressure initially rises in response to LH and then fluctuates before ovulation. The follicle thus enlarges greatly, its final size varying between species (e.g. bat 0.3 mm, rat 0.9 mm, pig 8 mm, man 25 mm, horse 30–50 mm).

During this final maturation of the follicle, the cells adjacent to the egg acquire their characteristic columnar shape. Initially the cumulus oophorus is attached to the membrana granulosa over a wide area (see Fig. 2.13). This zone gradually becomes reduced in extent, partly by movements and loosening of the granulosa cells but also by fluid accumulation. The area of attachment is eventually reduced to a small stalk, by which time the innermost layer of cumulus cells (the corona radiata) consists of long thin cells with nuclei remote from the egg. The oocyte with its investments of coronal cells may become free-floating in the follicular fluid shortly before ovulation.

These changes in the structure of the Graafian follicle are accompanied by a resumption of meiosis within the oocyte. At the onset of pre-ovulatory maturation, the egg is still a primary oocyte whose progression through meiosis was interrupted by a prolonged period of arrested development immediately following diplotene. At the time of the LH surge the oocyte contains lampbrush chromosomes, but the synthesis of RNA that is characteristic of early stages of growth is now terminated. The chromosomes subsequently shorten and thicken, and their lampbrush loops (Fig. 2.10) are withdrawn. Chiasmata move along the chromosomes to become terminal and the threads resemble 'crosses' and 'chains' within the nucleus (see Fig. 2.6). These events occur at the 'germinal vesicle' stage of meiosis (more correctly, diakinesis), and complete the prophase of the first meiotic division. Metaphase I rapidly ensues, in which the bivalents arrange themselves on (and become attached to) microtubules at the equator of the meiotic spindle. During anaphase the bivalents move to opposite ends of the spindle, which rotates gradually through 90°, so that the axis becomes radially orientated (see Figs. 2.6 and 2.16). The rotation is completed by early telophase when the surface of the egg near to the chromosomes may form a slight elevation. The repulsion of the two sets of chromosomes is now complete and the area of the spindle between them is elongated. Division of the cytoplasm of the oocyte rapidly occurs as a furrow around the more peripheral set of chromosomes. This division of the egg is unequal (non-equational) in that one daughter cell – the secondary oocyte – receives most of the cytoplasm while the other contains the minimum of ooplasm and is the first polar body (see Figs. 2.6 and 2.16). In some mammals this polar body soon degenerates, but in others it persists for prolonged periods, even through the first few cleavage stages. The first polar body sometimes divides into two slightly smaller daughter cells at a time when the second meiotic division occurs in the egg.

Shortly after the extrusion of the first polar body the secondary oocyte embarks on the second meiotic division. Prophase is very short or seemingly non-existent, and the chromosomes condense to form a half-moon-shaped mass at the periphery of the oocyte (chromatin mass stage). Spindle microtubules subsequently appear adjacent to the chromatin and

Fig. 2.16. Pre-ovulatory maturation of the egg. 1, germinal vesicle stage; 2, metaphase I; 3, anaphase I and rotation of the spindle; 4, polar body extrusion; 5, secondary oocyte at metaphase II, surrounded by a perivitelline space (PVS) between egg and zona pellucida (ZP). PB, first polar body; C, one of the cells of the corona radiata.

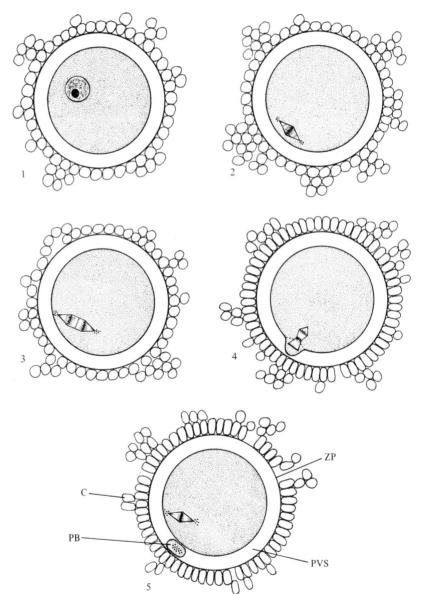

the chromosomes arrange themselves on the metaphase plate. Metaphase II in the majority of species is a stage of arrested development like the dictyate stage, and it is now that ovulation generally occurs. The subsequent meiotic maturation of the oocyte is dependent on penetration by a spermatozoon at fertilization, which is the subject of Chapter 6.

There can be no doubt that the resumption of meiotic changes within oocytes in Graafian follicles is induced by the LH surge, but the precise mechanism is open to dispute. If oocytes are removed from their follicles prior to the release of LH *in vivo* and placed in a chemically defined medium devoid of all hormones (both steroid and protein), they resume meiosis spontaneously. So we can infer that the Graafian follicle in some way inhibits the maturation of the oocyte until it is 'triggered' by LH. But this is not the whole story, since meiosis in mice can also be held up at two other stages: oocytes cultured in chemically defined media may fail to resume meiosis altogether (persistent germinal vesicle stage), and a proportion of the oocytes appears to be 'blocked' in meiosis at metaphase I. These oocytes are often smaller than those that complete the process, but otherwise are normal in appearance. There is some suggestion that oocytes whose maturation is 'blocked' at metaphase I are usually obtained from smaller follicles, and therefore that follicular size determines the potential for meiosis in the oocyte. Thus oocytes from pre-antral follicles will not respond to LH treatment prior to about stage 5–6 of growth, after which meiosis proceeds only to metaphase I. Only eggs from Graafian follicles can sustain meiosis to the point where ovulation normally occurs (metaphase II).

Many important maturational changes have to occur within the various components of the Graafian follicle before the normal surge of LH can exert its effect. It has recently been shown in the domestic pig that injections of human chorionic gonadotrophin (hCG) (which has an action similar to that of LH) on inappropriate days of the 21-day oestrous cycle, although inducing ovulation, result in infertility. The LH surge normally occurs on days 19–20 and at this time injections of hCG have little effect: the animals accept the boar and each of the ovulated eggs comes to contain one spermatozoon, which forms a male pronucleus (see Chapter 6). By contrast, when pigs are injected with the same amount of hCG on days 17–18 of the reproductive cycle they produce much lower quantities of oestrogen, will not accept the male, and ovulate immature oocytes. These oocytes have failed to resume meiosis (remaining at the germinal vesicle stage instead of reaching metaphase II) and become polyspermic after artificial insemination; in addition the spermatozoa cannot form male pronuclei (see Table 2.1 and Fig. 2.17). Thus LH (or hCG) can induce ovulation without affecting the meiotic status of the oocyte, or its ability to block polyspermy or induce sperm-head swelling. Furthermore, maturational changes have to occur in the follicular wall if sufficient oestrogen is to be produced to permit behavioural oestrus to occur.

The blockade of meiosis at the germinal vesicle and metaphase I stages could be due to immaturity of oocytes and/or follicle cells, or to changes in their metabolism. However, the change leading to resumption of meiosis is more likely to be mediated by steroid hormones, since these are known to accumulate in follicular fluid and their levels increase with LH treatment. The steroids presumably act on the cumulus cells rather than directly on the oocyte, possibly via metabolic intermediates. Clearly the follicle should be regarded as a dynamic structure in which the cellular components – granulosa, theca cells and oocyte – are interdependent. Thus the initial growth phase of the follicle probably occurs in response to factors from the oocyte influencing the granulosa cells.

### Ovulation

The oocyte is shed from the Graafian follicle by the process of ovulation at a precise time after the onset of the LH surge (see p. 34 and Table 2.2). The timing of ovulation in relation to the reproductive cycle varies considerably between species and may be dependent on diurnal rhythms, or induced by coitus (see Book 3, Chapter 6). Detecting the event of

Table 2.1. *Response of sexually mature pigs to injection of 500 IU of hCG on various days of the reproductive cycle*

| Day of hCG injection | Behavioural oestrus | Oestrogen in follicular fluid (ng/ml ± S.D.) | Mean no. ovulations | Primary oocytes (dictyate stage) (%) | Poly-spermic eggs (%) | Eggs with transformed sperm heads (%) |
|---|---|---|---|---|---|---|
| 17 | — | 75.3 ± 8.7 | 18.6 | 83.2 | 91.7 | < 10 |
| 18 | — |  | 12.4 | 17.4 | 38.0 |  |
| 19 | +/− |  | 12.9 | 2.1 | 2.4 |  |
| 20 | + | 263.8 ± 54.5 | 11.4 | 1.5 | 2.8 | > 90 |

*Source:* Data taken from R. H. F. Hunter, B. Cook and T. G. Baker, *Nature*, **260**, 150 (1976).

Fig. 2.17. Response of oocytes of sexually mature pigs to a single injection of gonadotrophin during the early or later follicular phases (days 17 and 20, respectively).

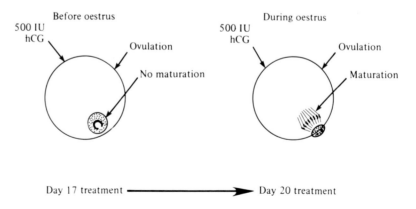

Day 17 treatment ⟶ Day 20 treatment

Table 2.2. *Variable features of ovulation in different mammals*

| Animals | Diameter of pre-ovulatory follicle (mm) | Interval between LH peak and ovulation (h) | Number of ovulations |
|---|---|---|---|
| Mouse | 0.55 | 12–15 | 8 |
| Hamster | 0.65 | 9–11 | 6 |
| Rat | 0.9 | 12–15 | 10 |
| Rabbit | 2.0 | 9–11 | 5 |
| Sheep | 8.0 | 25 | 1 |
| Pig | 8.0 | 40 | 10 |
| Rhesus monkey | 10.0 | 24–36 | 1 |
| Man | 25.0 | 28–36 | 1 |
| Cow | 15.0 | 40 | 1 |
| Viscacha (*Lagostomus maximus*) | — | — | 300–800 |

*Source*: From H. Peters and K. P. McNatty. *The Ovary*. Paul Elek; London, Toronto, Sydney and New York. (1980).

Fig. 2.18. The diagram on the left interprets the images shown on the right, which were obtained by ultra-sound scanning of a woman's abdomen shortly before the mid-point of her menstrual cycle. The size of the follicle is consistent with its being just pre-ovulatory. (Courtesy of Dr O. Djahanbakhch.)

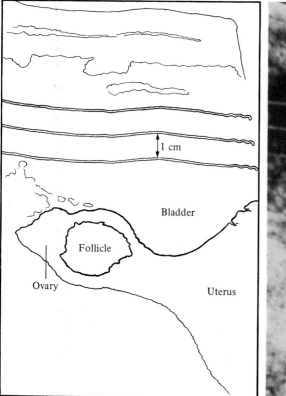
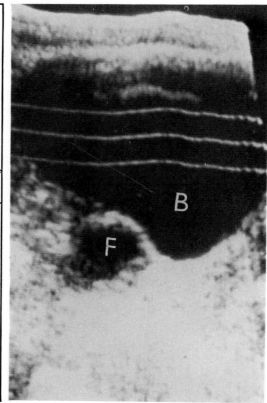

ovulation in the intact subject has long been the ambition of investigators, and now seems to be within our reach. Ultra-sound scanning has reached a sufficient state of precision to allow recognition of the human pre-ovulatory follicle and so with frequent scanning the moment of its rupture can be accurately timed (Fig. 2.18).

During the final phase of pre-ovulatory maturation, a small area of the wall of the Graafian follicle (and the overlying ovarian cortex) becomes thin and translucent, forming the 'stigma'. Studies of sectioned follicles have shown that thinning follows pyknosis of an area of the membrana granulosa, the cells being phagocytosed or merely shed into the follicular cavity. Translucence of the stigma may be partly due to its becoming devoid of blood capillaries. Towards the end of the period in which these changes occur, the stigma becomes raised and resembles a blister on the surface of the ovary. With the onset of ovulation the blister tears open and initially some watery follicular fluid spurts out. This is followed by a gentle oozing-out of a more viscid material containing the egg (in its cumulus oophorus), some granulosa cells, and a little blood from ruptured vessels. The somewhat gelatinous material containing the egg is 'wiped' off the surface of the ovary by the fimbria of the Fallopian tube, and carried by ciliary currents into the infundibulum – a process that is beautifully demonstrated in a colour movie produced by Richard Blandau at the University of Washington, Seattle.

So much for the observed process of ovulation, but what of the mechanism responsible for these changes? There can be little doubt that injections of either LH or hCG are effective in inducing ovulation, but the natural process *in vivo* probably involves a precise ratio of FSH:LH rather than LH alone. Several mechanisms have been proposed to account for the observed changes in the follicle wall, but most are easily discounted on physiological grounds. A combination of events may well occur, and considerable variation among species is only to be expected.

Ovulation has often been attributed to increase in pressure of follicular fluid, owing to either continuous secretion or contraction of the follicle by smooth muscle fibres and/or pressure exerted by the oviduct. But ovulation is not an explosive phenomenon, and manometric studies have shown that intra-follicular pressure merely oscillates about a value similar to that within capillaries. Thus pressure changes alone cannot account for the release of oocytes in most species, and injection of excess fluid into follicles of the pig does not induce ovulation.

The most widely accepted theory is based on the idea that the blister-like stigma results from the closing down locally of capillaries (ischaemia). The blood flow into the ovary increases following treatment with LH and also with histamine, but is unaffected by FSH, prolactin or adrenalin. Ischaemia in the region of the stigma is believed to be induced by the combined effects of high blood pressure and forces applied between the follicle and ovarian cortex. Necrotic changes in the membrana granulosa resulting in the translucence of the stigma could be due to the action of enzymes, a number of which have been found in follicular fluid (Table 2.3). Injections of pronase, collagenase or nagrase into rabbit follicles have been shown to induce ovulation in 62–100 per cent of Graafian follicles. Trypsin is less effective, and chymotrypsin, hyaluronidase, lysozyme and saline solution are ineffective. Apparently the proteolytic enzymes are produced by follicular cells in which the relevant proteins can be detected histochemically. Furthermore, injections of actinomycin D or puromycin (substances that block protein synthesis) into the ovaries of rabbits inhibit ovulation. The enzymes presumably have only a local effect since few granulosa cells become pyknotic. Intra-follicular pressure initially rises with enzyme treatment but then subsides to normal levels before ovulation occurs.

*The egg at ovulation*

Ovulation occurs in most mammals when the oocyte has reached the metaphase of the second meiotic division. There are, however, exceptions to this general rule. Thus in the fox and dog ovulation occurs around metaphase I and the first polar body is extruded shortly thereafter. By contrast the second meiotic division is said to be completed prior to ovulation in certain primitive insectivores (the Centetidae), and fertilization begins within the Graafian follicle.

Table 2.3. *Enzymes in human, bovine and porcine follicular fluid*

| | Presence in different animals | | |
|---|---|---|---|
| Enzyme | Man | Cow | Pig |
| Endopeptidase | + | − | + |
| Proteinase | − | + | + |
| Plasmin | − | + | − |
| Aminopeptidase (cytosol) | + | − | + |
| Dipeptidase | − | − | − |
| Alkaline phosphatase | + | + | + |
| Adenosine triphosphatase | − | − | − |
| Acid phosphatase | + | + | − |
| Fructose bisphosphate aldolase | + | − | + |
| Lactate dehydrogenase | + | + | + |
| Aspartate aminotransferase | + | + | + |
| Alanine aminotransferase | − | + | + |
| Collagenase | − | − | + |
| Hyaluronoglucosidase | + | + | − |
| Pyrophosphatase | + | − | − |

+, presence of an enzyme; −, enzyme not detected.
*Source*: From R. E. Jones. *The Vertebrate Ovary*. Plenum Press, New York and London (1978).

*Fate of the follicle after ovulation*

Dramatic changes occur in the follicle after ovulation. They result in the formation of a true endocrine gland, the corpus luteum, within the cortex of the ovary (see Fig. 2.4). The follicle initially collapses to a fraction of its former size and the membrana granulosa is thrown into folds. The remnants of the antrum are rapidly obliterated by proliferation of granulosa cells which become transformed into lutein cells; there is also an infiltration with capillaries from the theca interna. Some of the thecal cells pass in with the capillaries to form subdividing walls, the trabeculae, but the main component of the gland is derived from the granulosa. The lutein cells contain secretory granules and in the cow they also contain the yellow pigment from which the name corpus luteum is derived (i.e. 'yellow body').

The function of the corpus luteum is to secrete a steroid hormone, progesterone, which is important in controlling the length of the reproductive cycle in many species and also for maintaining pregnancy (see Book 3, Chapter 7). The gland has a finite life depending on whether or not pregnancy ensues. Towards the end of its life, the secretion of progesterone ceases and the lutein cells degenerate. Regression is completed by an invasion of fibroblasts which convert the gland into a mass of scar tissue (corpus albicans; Fig. 2.4).

**Atresia**

We have already remarked that few oocytes and follicles survive to the stage of ovulation, the great majority being eliminated by a degenerative process known as atresia. By far the greatest 'wastage' of germ cells occurs before birth, and affects oogonia at interphase and during the course of mitotic division, as well as oocytes at all stages of meiotic prophase (especially pachytene and diplotene). In the human ovary, for example, some five million oogonia and oocytes are eliminated from the ovaries between the fifth month of gestation and full term. The factors responsible for such waves of atresia are obscure, although genetic defects and errors in metabolism are probably involved.

That genetic defects are important in the induction of atresia is shown from studies of mutant mice. Mutations at the *W* locus appear to induce degeneration of germ cells shortly after they have colonized the genital ridges. Similarly, human females devoid of one X-chromosome per cell (XO or Turner's syndrome) are characterized by the absence of oocytes after birth. The ovaries of fetuses with this genetic constitution are initially normal but the germ cells are eliminated before puberty. Turner's syndrome in women is thus accompanied by total and permanent sterility. The condition is less deleterious in female mice in which XO germ cells not only survive but can give rise to viable offspring (see Book 2, Chapter 3).

Circumstantial evidence for a genetic influence on pre-natal atresia derives from the observation that morphological changes affect the chromosomes long before cytoplasmic defects can be detected. But if genetic defects were the only cause of atresia, this would imply that the load of mutations and chromosomal aberrations would be enormous; therefore it is unlikely to be true. Other errors in the germ cells, attributable to defects in metabolism or shortage of nutrients (due to inadequate follicular cells or paucity of blood supply) are probably important contributing factors.

Atresia continues to deplete the population of oocytes in the ovaries of post-natal mammals. The population declines rapidly during pre-pubertal life and then at a lower rate throughout adult life; these points were discussed early in this chapter. The rate of decline in the population of oocytes varies considerably among species and even among strains within a single species. In CBA mice, for example, the number of oocytes reaches zero when the animals are about 300–400 days old, whereas mice of the RIII and A strains retain some 1000 oocytes in their ovaries at the same age. Significantly, hybrids of the CBA strain and A stock have a higher population of follicles during middle to old age than either of the parent groups.

The earliest signs of atresia in 'growing' and Graafian follicles are: (*a*) condensation of chromosomes and wrinkling of the nuclear envelope in the oocyte, and (*b*) pyknosis of the nuclei of granulosa cells, which are then detached from the membrana granulosa and become free-floating in the follicular fluid. The oocyte may appear to complete its meiotic divisions abnormally ('pseudomaturation') or undergoes fragmentation so that it superficially resembles an embryo during cleavage. Pseudomaturation changes are occasionally observed in atretic primordial follicles, the granulosa cells of which usually survive while the oocyte is phagocytosed by macrophages. There can be no doubt that the rate of atresia is controlled by gonadotrophic hormones; after unilateral ovariectomy the level of circulating FSH in mice is seemingly unaffected and the remaining ovary undergoes compensatory hypertrophy so that as many 'growing' and Graafian follicles occur as in the two ovaries of controls. The rate of ovulation for the ovary doubles and the number of offspring per litter is unaffected. But the reproductive life-span and *total* number of offspring born to these mice are reduced, since no 'new' oocytes can be formed to replace those in the extirpated ovary (these features are discussed more fully in Book 4, Chapter 7).

The rate of decline in the oocyte population with increasing age is considerably reduced, although not altogether prevented, by hypophysectomy. Transplantation into intact mice of the ovaries from those that were hypophysectomized 300 days previously leads to a resumption of follicular growth and oocyte depletion by atresia. Injections of the synthetic oestrogen stilboestrol into hypophysectomized immature rats retards atresia affecting growing follicles, whereas treatment with exogenous gonadotrophins (such as PMSG) hastens follicular atresia. The effect of

PMSG varies with increasing dose from a follicle-stimulating effect, to atresia, and then to luteinization.

Exposure of the ovaries to ionizing radiation has the opposite effect to hypophysectomy: the rate at which oocytes are eliminated is initially increased. But the process of atresia following irradiation is not necessarily the same as that occurring spontaneously in oocytes. Radiation induces chromosomal breaks which may or may not undergo repair. The damaged cells are eliminated within hours or days of treatment, whereas those that are unaffected or have undergone repair may persist for prolonged periods, to be ovulated and fertilized. The dose of radiation required to destroy a given population of oocytes varies considerably among species and also depends upon age. Primordial oocytes in rats and mice are highly sensitive, while those in guinea pigs, rhesus monkeys and the human female, are fairly resistant to X-irradiation. The results of a recent study indicate that the control of radiation-induced atresia by hormone treatment may be similar to that occurring spontaneously. We have at present no idea of the nature of atresia, nor how long is needed for an affected cell to be eliminated from the ovary. Grossly abnormal cells may undergo lysis *in situ* ('pools' of degeneration), or be phagocytosed by surrounding somatic cells. Clearly the process of atresia is regulated by gonadotrophins and probably steroids as well, although the precise controlling mechanisms are obscure. One of the most fundamental problems in reproductive biology that still remains completely unsolved is why one oocyte should be selected to undergo meiosis and ovulation, while its immediate neighbours suffer atresia and are eliminated from the ovary. Furthermore, why should the granulosa envelope of one particular oocyte be stimulated to undergo mitosis and growth while other apparently identical cells are unaffected? These forms of 'germinal selection' do not seem to operate in the male and cannot be due to genetic faults alone since the implied mutation frequency would be enormous. New techniques need to be devised if further studies of the processes of atresia and germinal selection are to enhance our knowledge of such fundamental issues.

A great deal of interest among biologists is stirred by the growth and differentiation of the oocyte. This is understandable, for not only are oocytes the largest of the body's cells but they also have the potential to develop into complete new individuals. They may even achieve this feat unaided by the spermatozoon – as we shall see in the next chapter, mammalian eggs can at least begin to develop parthenogenetically – and so must carry within themselves all the necessary machinery. Fittingly enough, oogenesis includes meiosis, a complex process that bestows uniqueness on the individual.

## Suggested further reading

A quantitative and cytological study of germ cells in human ovaries.
T. G. Baker. *Proceedings of the Royal Society of London, Series B* **158**,
417–33 (1963).

Electron microscopy of the primary and secondary oocyte. T. G. Baker. In
*Advances in the Biosciences* 6, pp. 7–23. Ed. G. Raspé. Schering Symposium
on Intrinsic and Extrinsic Factors in Early Mammalian Development, Venice,
1970. Vieweg-Pergamon (1971).

The fine structure of oogonia and oocytes in human ovaries. T. G. Baker and
L. L. Franchi. *Journal of Cell Science* **2**, 213–24 (1967).

The uptake of tritiated uridine and phenylalanine by the ovaries of rats and
monkeys. T. G. Baker, H. M. Beaumont and L. L. Franchi. *Journal of Cell
Science* **4**, 655–75 (1969).

Mammalian eggs in the laboratory. R. G. Edwards. *Scientific American* **215**
(August), 72–81 (1966).

*Marshall's Physiology of Reproduction*, 3rd edn. Ed. A. S. Parkes. London;
Longmans (1956). (Especially vol. 1, part 1, chapter 5 by F. W. R. Brambell.)

*Reproductive Physiology of Vertebrates*. A. van Tienhoven. Saunders;
Philadelphia (1968).

*The Chromosomes*, 6th edn. M. J. D. White. Chapman and Hall; London
(1973).

*The Ovary*, 2nd edn, 3 vols. S. Zuckerman and B. J. Weir. Academic Press;
London and New York (1979). Also chapter 4 on Atresia by D. L. Ingram in
the first edition (1962).

*Fundamentals of Obstetrics and Gynaecology*, 2 vols. D. Llewellyn Jones. Faber
and Faber; London (1973).

*The Developing Human: Clinically Orientated Embryology*. K. L. Moore.
Saunders; Philadelphia (1973).

*The Ovary*. H. Peters and K. P. McNatty. Paul Elek; London, Toronto, Sydney
and New York (1980).

*The Vertebrate Ovary*. Ed. R. E. Jones. Plenum Press; New York and London
(1978).

# 3

# The Egg

*C. R. AUSTIN*

By 'egg' we mean the entity that begins as a single cell and is released from the ovarian follicle at ovulation, later to become fertilized and develop into an embryo, but since all this involves a series of events, during which the original cell progressively divides into several, it is natural that more specific names have been applied at different stages. What is ovulated in most animals is a secondary oocyte, with the first polar body emitted and the chromosomes arranged on the second maturation spindle (as described in Chapter 2). Exceptions to this most frequent course of events are provided by the eggs of the dog and fox which are ovulated while still primary oocytes, i.e. containing an intact germinal vesicle or a stage shortly after the start of the first meiotic division. In these animals, the first meiotic division and emission of the first polar body take place wholly or mostly in the Fallopian tube. If mating has occurred, spermatozoa may penetrate eggs at any time during these events but the sperm head remains almost unchanged until emission of the second polar body. In the majority of mammals the second meiotic division is arrested at metaphase and normally proceeds only after sperm entry. When the second polar body emerges the egg becomes strictly an ootid, though the term is rarely used; shortly afterwards, maternal and paternal chromosomes are united for the first cleavage metaphase and the cell is called a zygote, implying that the hereditary contributions of the parents are now yoked to work together. After this we find people prepared to talk as readily about 2-cell, 4-cell and more-cell 'eggs' as about cleavage embryos, morulas and blastocysts. For some, especially those with a medical background, even the implanted embryo, while still a relatively small spherical structure bounded by the

Fig. 3.1. Eggs of several mammals, drawn to about the same scale. 1. Echidna, a monotreme. Fertilization is in progress and the two pronuclei are visible. The egg is enclosed by a thin zona pellucida which is covered by a modest layer of albumin, a shell membrane and a soft shell. 2. Australian native cat, a marsupial. This is a mature oocyte with the first polar body and the second maturation spindle. The cytoplasm partly surrounds a large yolk body. Beyond the thin zona pellucida are several layers of cells, the remnants of the cumulus oophorus. 3. The two-cell egg or early embryo of the Virginia opossum, another marsupial. This, too, is contained within a thin zona pellucida. Externally there is a broad layer of mucin, limited by a shell membrane.

Small cytoplasmic or yolky extrusions in the perivitelline space are regularly seen in these eggs, but their significance is not known. 4–10. The eggs of eutherian mammals: dog, man, field vole, sheep, noctule bat, guinea pig and rabbit, in that order. These eggs are distinguished by having a relatively thick zona pellucida. Retention of cumulus cells through early cleavage is a feature of dog and cat eggs. Rabbit (and hare) eggs, alone among the eggs of eutherian mammals, accumulate a very wide layer of mucin as they traverse the Fallopian tube. A, albumin; C, cumulus cells; M, mucin; S, shell; Sm, shell membrane; Y, yolk; Zp, Zona pellucida. (1, 2, 3 and 8 are based on histological sections or fixed eggs, the remainder on living eggs.)

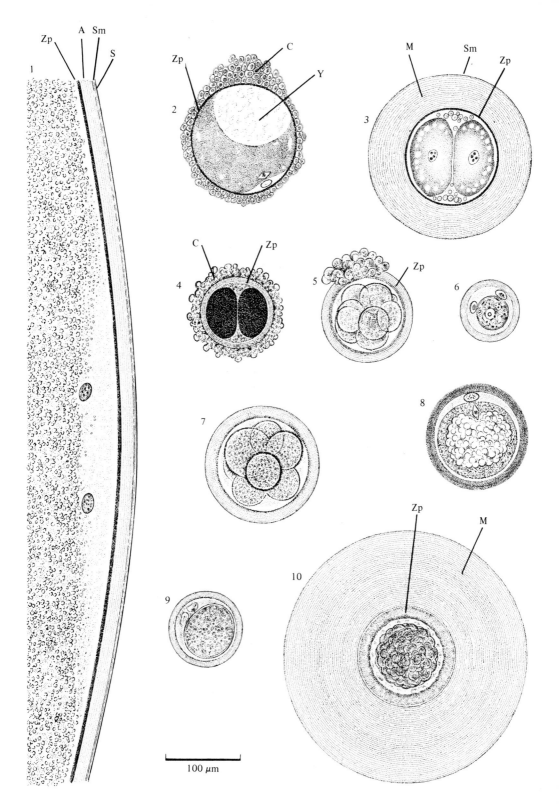

100 μm

chorion, is an egg or an ovum. In this chapter, however, we will restrict our attention mostly to the secondary oocyte, the ootid and the zygote.

**General structure of the egg**

The egg of the eutherian mammal consists of a spherical mass of cytoplasm, 70–120 $\mu$m in diameter – or of two or more cells into which this mass has divided – bounded by a plasma membrane and contained within a glycoprotein envelope, the zona pellucida (Fig. 3.1). The zona pellucida is peculiar to mammalian eggs in its structure, composition and presumed functions, though the eggs of many non-mammals have similar-looking investments. Between the cell surface and the envelope is a fluid-filled space, the perivitelline space. The use of this name persists though it acknowledges the word 'vitellus' to denote the cytoplasmic body of the egg; vitellus is a temptingly convenient term but strictly misleading here, because it was traditionally used to denote the yolk of the hen's egg which is very largely the nutrient reserve. (As the monotreme egg is also very yolky, the term would be appropriate for it too.) The amount of nutrient in eutherian mammalian eggs is minute, and the bulk of the cytoplasmic body is destined eventually to form both fetus and fetal membranes. Probably 'ooplasm' is the best term for the living part of the mammalian egg. Beyond the zona pellucida, the freshly ovulated eggs of marsupials and eutherian mammals are invested by a broad layer of granulosa cells, the cumulus oophorus (Fig. 3.2).

When laid, the eggs of monotreme mammals, the duck-billed platypus *Ornithorhynchus* and the spiny anteater or echidna *Tachyglossus* are quite large (a centimetre or so in diameter) and resemble hen eggs in having a large mass of yolk, an albuminous layer or 'white', and a shell membrane, and they are covered externally with a soft shell. (An indication of the size

Fig. 3.2. A freshly ovulated rat egg, showing the surrounding mass of cumulus cells embedded in a gel matrix.

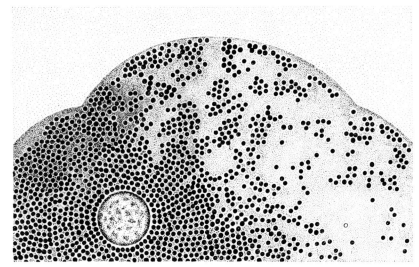

is tellingly given by Fig. 2.8, Book 6 of the First Edition.) The yolk is limited by a plasma or vitelline membrane and a thin zona pellucida; at one point on the surface of the yolk lies a small aggregate of cytoplasm containing the egg nucleus and organelles. As in eutherian mammals, the egg becomes fertilizable after emission of the first polar body, when the egg chromosomes are arranged on the second maturation spindle.

Marsupial eggs are much more like those of eutherian mammals than monotremes; they range up to about 250 μm in diameter, have a modicum of yolk, which may be gathered into a single mass, and a moderately developed zona pellucida. Externally, however, they do acquire a mucoid coat in the oviduct, which is less generous than in the monotremes, as well as a shell membrane, but no shell.

Some eutherian mammalian eggs, namely those of the Lagomorpha, the rabbits and hares, also receive a mucoid coat as they pass along the oviduct and this gets quite thick by the time they reach the utero-tubal junction, but there is no sign of a shell membrane.

## Cytoplasm

The cytoplasm of the egg is bounded by the plasma membrane and contains the nucleus or, during periods of division, the condensed chromosomes, as well as a variety of organelles, persisting structures which seem to have several different functions. In addition, there are inclusions of various kinds, ranging from the fairly well characterized cortical 'granules' (cortical vesicles would be a better term) and microtubules and microfilaments of the cytoskeletal system, to a heterogeneous array of vesicles, droplets and particles which in eutherian mammals are thought to represent the nutritive reserve – a much reduced form of yolk. The distinctive species differences that exist in the appearance of eggs stem mainly from the size, form and density of cytoplasmic organelles and inclusions. (The possible evolutionary significance of differences in egg morphology is discussed in Book 6 of the First Edition, Chapter 5.)

The main organelles in eggs are the mitochondria, the endoplasmic reticulum and the Golgi system. These organelles contain an array of enzymes and are involved in important metabolic and synthetic activities. In somatic cells mitochondria commonly take the form of elongated double-walled sacs with incomplete partitions or cristae, formed by the infolding of the inner membrane. In oocytes and 2-cell eggs the mitochondria are more spherical in form, and the cristae are few in number, tending to be flattened against the wall of the mitochondrion, so that superficially it often looks as if there were none. As the egg (embryo) goes through successive cleavage divisions after fertilization, the mitochondria develop more cristae and become similar to somatic-cell mitochondria. Concurrently, the metabolic capability of the system increases. Endoplasmic reticulum in oocytes takes the form of relatively sparse, smooth-walled, flattened vesicles; it is thought to be involved in a low level of protein and nucleic acid synthesis. During cleavage the system becomes more complex, and rough endoplasmic reticulum (i.e. armed with ribosomes) comes to be a feature of secretory tissues as these differentiate during embryonic development. In the ovary, the oocyte Golgi is implicated specifically in vitellogenesis or yolk synthesis and seems also to be responsible for the formation of the cortical granules.

Cortical granules evidently play an important part in the egg's response to sperm penetration (Fig. 3.3), essentially as part of the mechanism preventing the entry of more than one spermatozoon (which would lead to anomalous development in eutherian and marsupial embryos, but probably not in monotreme embryos because large eggs like those of monotremes may well have physiological polyspermy). This function is a feature of the block to polyspermy which is discussed in Chapter 6.

Fig. 3.3. Cortical granules in a rat egg before and after exocytosis has resulted in release of their contents into the perivitelline space. × 10 000 (Based on electron micrographs of D. G. Szollosi.)

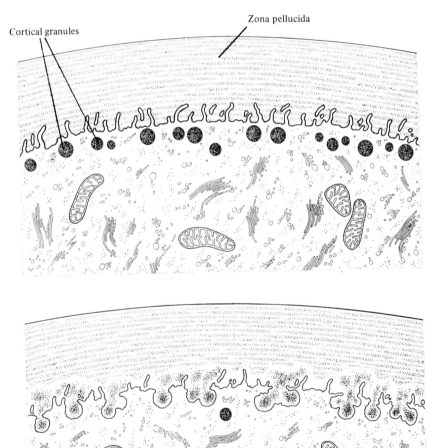

Microtubules are involved in the construction of the meiotic and mitotic spindles. Spindles usually have centrioles at their poles but Dan Szollosi and his colleagues have shown with the aid of highly refined electron microscopy that in the mouse the meiotic spindles (and even the mitotic spindles of the first few cleavage divisions) lack poles altogether, and thus are barrel-shaped (Fig. 3.4). The microtubules are marshalled, not by centrioles but by a number of well defined organizing centres which become arranged in a circle at each end of the spindle.

Microfilaments appear to exist regularly in the cortical region of the egg and of the blastomeres into which the zygote divides during cleavage; their main function in life could well be that of providing the means whereby the cytoplasmic body of the egg is coerced to yield a polar body or to split into two equal parts as in cleavage.

The plasma membrane of the egg is broadly similar to the plasma membranes of other cells; it evidently has its own peculiarities of structure and function, but these are at present poorly defined. Accordingly we must speak in general terms about this most important structure, and infer that a great deal of what is known about the membranes of other cells could be true also for those of eggs.

In essence, plasma membranes consist of a double layer of phospholipid molecules, oriented so that their fat-soluble ends are pointing towards each other and their water-soluble ends presented away (Fig. 3.5). Strong

Fig. 3.4. The barrel-shaped spindle seen characteristically in mouse eggs during the meiotic divisions and the first few cleavage divisions. Homologous pairs of chromosomes are arranged around the equator of the spindle. Nothing resembling centrioles can be found at the poles, though these structures seem to be essential components of the division apparatus in most cells. Instead, there is a ring of microtubule-organizing centres at each end. (After Szollosi, Calarco and Donahue (1972) – see Suggested Further Reading.)

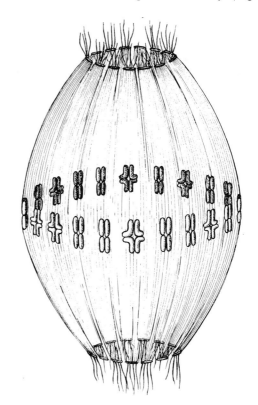

intermolecular forces preserve the integrity of this bimolecular lipid leaflet. In addition the hydrophobic region in the middle exerts a powerful integrating influence. 'Floating', as it were, in this phospholipid 'sea' are protein 'icebergs', molecular aggregates (though depicted as individual molecules in Fig. 3.5) which may exist in only one layer or extend through both and thus constitute a connection between the cell's inside and its environment. Those connecting protein particles having hydrophilic cores probably provide channels for the passage of ions; since control of ion movement is obviously necessary, we recognize the existence of 'gated' channels (specifically for sodium and calcium transport). As we shall see in Chapter 6, calcium ions play a crucial role in the acrosome reaction and the cortical granule reaction, and so study of the means for control of calcium ion movement has a high priority. Adsorbed proteins (suggested on the right of Fig. 3.5) can be detached relatively easily, and this may be the nature of the material whose removal seems to be a major element in sperm capacitation (see Chapter 6).

Fig. 3.5. A diagram showing the possible molecular arrangement of a plasma membrane. Most of the structure consists of phospholipids, arranged in two layers, with the fat-soluble ends of the molecules lying in the hydrophobic zone in the middle. Steroids also exist in these layers. At intervals protein molecules are held in the membrane, their fat-soluble portions being in the hydrophobic zone and their electrostatically charged water-soluble extremities extending either into the cell cytoplasm (below) or the external medium (above). Some proteins are not integral with the membrane but simply adsorbed onto the surface, as on the right. A single protein traversing the membrane (as on the left) may link up with the cytoskeleton of the cell and serve to transmit influences from the external environment to the cytoplasm and so influence cell movement, or, if the protein has a water-soluble core, it could provide a channel for the passage of ions into and out of the cell. Many membrane proteins are glycoproteins (as in the middle), the name signifying that they carry an array of saccharide molecules at their external ends; such proteins seem especially important in cell–cell recognition and attachment – processes of high specificity. Proteins may be restricted to one or other phospholipid layers (middle and right); two such proteins may become associated (as in the middle) and function in systems where the protein in the outer (upper) layer is a specific receptor and the one in the inner (lower) layer acts as a 'second messenger', such as adenylate cyclase, transmitting the influence to the cell cytoplasm.

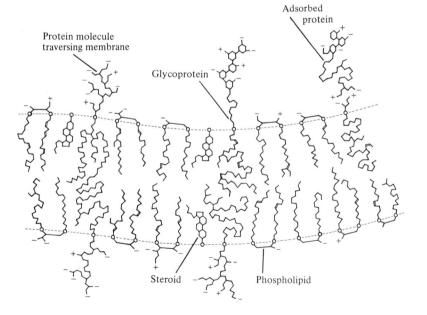

The floating protein particles are also important in at least two other ways. Proteins characteristically have fat-soluble and water-soluble ends, the water-soluble components being notably the straight or branched chains of sugar molecules that form the distinctive parts of glycoproteins. These sugar molecules can be of several different kinds and can be connected together in several different ways; the permutations of this arrangement make possible several hundred (perhaps thousand) different glycoproteins, providing an incomparable device for imbuing cell surfaces with distinctive signatures. We have many reasons for believing that the specificity of cell-to-cell reactions – like the sperm–egg specificity that tends to prevent hybrid fertilization – depends on just this property. In addition, the sugar residues are not only water-soluble but also they carry electrostatic charges; the glycoside terminals of glycoproteins represent the main location of the forces of repulsion between cells (helping to prevent, for instance, the re-fusion of oocyte and polar body or of the blastomeres in cleaving embryos) or between membranes in the same cell (as between the plasma and outer acrosome membrane in spermatozoa – relevant to the acrosome reaction, which is discussed in Chapter 6).

## Nuclear elements

On the evidence we have at present, it is a reasonable assertion that all mammalian oocytes stop at the metaphase of the second meiotic division until (most mammals) or unless (dog and fox) sperm penetration has occurred, so that condensed chromosomes represent the state of the nucleus in oocytes immediately before fertilization. When the spermatozoon enters the egg its nucleus is also in a condensed state (hypercondensed, in fact), made possible apparently by incorporation of cysteine-rich protamines in the nuclear protein. The first part of fertilization, then, consists of decondensation, and the formation of vesicular pronuclei with their complement of distinctive refractile nucleoli. During syngamy, late in fertilization, chromosomes condense again, facilitating their assembly together (maternal with paternal) on the first cleavage spindle. Throughout cleavage one sees a succession of similar condensations of chromosomes, and formation of vesicular nuclei, followed by decondensation, disappearance of nuclear envelopes and organization of mitotic spindles.

The striking-looking pronuclear nucleoli are, it seems, peculiar to pronuclei in that they contain little if any nucleic acid, have no discernible associated chromatin, and tend to vary a good deal in numbers, both between species and at different phases of fertilization in the same species. This is in line with the lack of RNA synthesis during fertilization. Progressively during cleavage the nucleolar system comes to assume the characters of that seen in somatic cells, and concomitantly the rate and variety of RNA synthesis increases, reaching high levels at the blastocyst stage. Messenger RNAs become detectable at the 4-cell stage in mouse embryos, but most of the RNA synthesized throughout is ribosomal RNA.

Evidence for the transcription of the embryonic genome begins to appear at the two-cell stage; in this particular, frog and sea urchin embryos differ, for with them early development, to about the blastula stage, is dependent on 'long-lived' maternal messenger RNA.

### Zona pellucida

The degree of development of this investment provides a clear-cut distinction between the eggs of monotremes, marsupials and eutherian mammals. The properties of this structure have been best studied in eutherian mammalian eggs, but the available evidence suggests that they are similar in the eggs of the other mammalian classes. The zona is made of glycoprotein and Bonnie Dunbar and her colleagues have recently shown that, in the pig egg, this consists of about 71% protein and 19% carbohydrate. Notable among the sugars identified were fucose, mannose, galactose, $N$-acetylgalactosamine and $N$-acetylglucosamine; sialic acid in glycosidic linkage, and sulphate and phosphate esters were also detected as true components of the zona. The solubility of the zona in varying conditions of pH, ionic concentration, temperature and the presence of certain detergents and urea led to the conclusion that the component glycoproteins were organized as supramolecular complexes held together by non-covalent forces. Enlarging on this aspect, Jeffrey Bleil and Paul Wassarman announced that they had shown the mouse zona to have three

Fig. 3.6. Drawings from scanning electron micrographs of the outside (left) and inside (right) of the hamster zona pellucida. (Based on electron micrographs by D. M. Phillips and R. M. Shalgi. *J. Exp. Zool.* **213**, 1 (1980).)

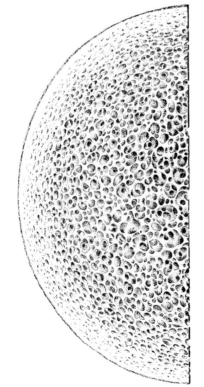
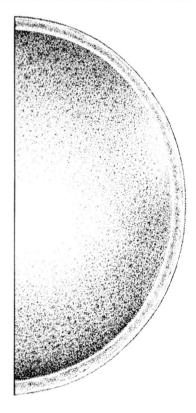

main glycoproteins which they identify as ZP 1, 2 and 3, with molecular weights of 200 000, 120 000 and 83 000, respectively. They found that ZP 3 interfered with sperm–egg binding and so this protein could function as the sperm receptor on the zona surface (and more will be said on this point in Chapter 6).

Recent observations by David Phillips and Ruth Shalgi with the scanning electron microscope show the hamster zona pellucida to have a remarkable appearance, under the conditions required for that study (Fig. 3.6). It seems to be made up of several layers of material, each perforated very extensively, so as somewhat to resemble an old-fashioned bath sponge. When the inside of the zona was examined it looked quite different, since this surface lacked obvious perforations but appeared to be thrown up into a vast number of minute prominences (Fig. 3.6). Interpretation is not particularly easy as the fixation and dehydration of this normally highly hydrated structure must produce major artefacts; in addition, the hyaluronidase treatment that the eggs received to free them of follicle cells may have added to the final effect, since earlier work has indicated that some part of the zona is digested by this enzyme. Certainly the normal zona pellucida does not have holes all over the surface and, in the newly ovulated egg, it appears to be impermeable to molecules much smaller than proteins. Later the permeability increases, so that proteins can eventually pass through. Virus particles are known to get through too, but that probably depends on the action of viral enzymes.

Between the zona pellucida and the cytoplasmic body of the egg is the fluid-filled perivitelline space. Before the meiotic divisions begin, this is simply a potential space but, with the emission of each polar body, fluid is evidently released from the egg making the space real. In some animals,

Fig. 3.7. A rabbit egg (morula) after the sixth or seventh cleavage division. The numerous spermatozoa in the perivitelline space and zona pellucida are quite a common sight in rabbit eggs. Around the zona pellucida, part of the mucin coat is evident.

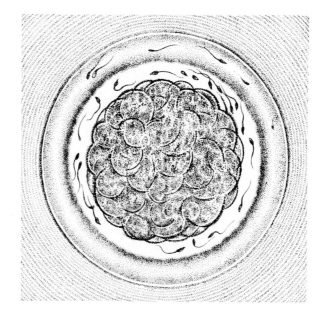

like the hamster, there is a further release of fluid in cleavage, so that embryos of eight-cell and later stages have very wide perivitelline spaces. In the rabbit, supernumerary spermatozoa, prevented from fusing with the egg by the block to polyspermy, but able to pass through the zona pellucida because of the absence of a zona reaction in this animal, can be seen in the perivitelline space, often in large numbers (Fig. 3.7).

**Cumulus oophorus**
The structure of the cumulus oophorus and its derivation from the granulosa cells of the follicle have already been described in Chapter 2. During its life in the follicle the oocyte evidently gets much of its nurture through the mediation of the cumulus cells; junctional complexes of probable transfer function are seen at the contact points between the cytoplasmic processes of these cells and the oocyte. Around the time of ovulation the cytoplasmic processes are withdrawn or break down, and so one looks for another function for the cells in those animals in which the cumulus persists. In all mammalian species so far investigated, except for the opossum *Didelphis virginiana* and the phalanger *Trichosurus vulpecula*, the egg arrives at the site of fertilization in the ampulla of the Fallopian tube still surrounded by the cumulus cell mass. Where it persists, we are tempted to infer that the cumulus provides a suitable surface against which the fimbrial cilia can work to move the egg into the Fallopian tube. This action is strongly supported in the film produced by Richard Blandau, but that leaves us with the problem of means of transport for the opossum egg. (Egg transport, however, is really the business of a later chapter, so we should leave it there.)

An additional possible use for the cumulus is that its radially arranged cells help to guide the fertilizing spermatozoon towards the egg, but this cannot hold for the cow and sheep, in which the cumulus breaks down and the eggs become denuded before sperm penetration. In the rabbit and rodents, the cumulus can persist for about as long as the egg remains fertilizable, but breaks up rapidly if mating has occurred and spermatozoa are present at the site of fertilization. In the dog and cat, the cumulus partly disintegrates after sperm penetration, but a layer or two of cells closest to the egg remain attached for much longer, through the first few cleavage divisions of the embryo. With all these variations in behaviour, we are hard put to it to settle upon a likely function for the cumulus, once the follicular oocyte is fully grown.

Break-up of the cumulus cell mass is brought about by dissolution of the gel matrix and migration of the cumulus cells. Dissolution may be caused by a release of protease – Richard Mumford and his colleagues have detected several proteinase activities in cumulus cells. When spermatozoa are about, there may be sufficient free hyaluronidase and acrosin in the vicinity to accelerate cumulus disintegration; this certainly seems to be true in rodents in which disintegration is much quicker if mating has taken place.

## Polar bodies

In Chapter 2 we saw how and when polar bodies are formed; it remains to say a few words about their number and persistence. In many invertebrates the first polar body divides into two shortly after its formation, and these are joined by the second polar body to make a total of three, which commonly persists throughout cleavage. (The second polar body cannot normally divide, being already haploid.) In mammals, various things can happen with the first polar body: it may undergo division (rare, but reported occasionally), remain undivided but persist (rabbit, hamster) or break up and disappear early (rat, some mouse strains). Thus the presence of two polar bodies can signify sperm entry and the completion of the second meiotic division, but if there is only one polar body this does not necessarily mean that fertilization has not been initiated. Then again, spontaneous activation may lead sometimes to a degree of parthenogenetic development, and the second polar body is often (but not invariably) emitted, so that the presence of two polar bodies is not a reliable indication of sperm entry.

In the mouse, one effect of ageing in the ovulated but unfertilized oocyte is the migration of the second meiotic spindle to the centre of the egg; when the division occurs, the egg is cleaved into two cells of similar size, one of which represents the polar body and the other the ootid. This form of development initiation is referred to as 'immediate cleavage' and we will return to it later in Chapter 6.

## Fertile life

By 'fertile life' of the egg we mean the period after ovulation during which the egg remains capable of fusing with a spermatozoon and giving rise to

Table 3.1. *Estimated fertile life of some mammalian eggs*

| Animal | Length of fertile life |
| --- | --- |
| Rabbit | 6–8 hours |
| Guinea pig | Not more than 20 hours |
| Rat | 12–14 hours |
| Mouse | 8–12 hours |
| Hamster | 5 hours, 12 hours |
| Ferret | Up to about 36 hours |
| Pig | About 20 hours |
| Sheep | 15–24 hours |
| Cow | 22–24 hours |
| Mare | About 24 hours |
| Rhesus monkey | Probably less than 24 hours |
| Man | Not more than 24 hours |

*Source*: From C. R. Austin. Fertilization. In *Concepts of Development*. Ed. J. Lash and J. R. Whittaker. Sinauer Associates; Stamford, Connecticut (1974).

a normal embryo that develops through to birth. Approximate estimates for various mammals are given in Table 3.1, assessed on the basis of successful pregnancies. Generally speaking, eggs remain fully fertile for only about a day after ovulation, but in some animals for only half a day. If the delay between ovulation and fertilization is longer than the fertile life, embryonic development may be initiated but is likely to be abnormal and of limited duration (Table 3.2). (This point is taken up again in Book 2, Chapter 5.) Thus eggs remain fertilizable for longer than they remain capable of giving rise to normal embryos, and surprisingly enough the deterioration that shows up in anomalous embryonic development can occur in some eggs well before the end of the oestrous or receptive period of the female. A certain proportion of embryonic or 'prenatal' loss is therefore likely even under natural conditions, being especially noticeable in animals, like pigs, that bear litters.

Further analysis has shown that, as a result of ageing, the capacity of eggs to undergo the whole process of fertilization is lost before the capacity to fuse with a spermatozoon. Ageing is thus a progressive change in two respects: it affects some eggs earlier than others, and some processes (such as pronuclear growth) earlier than others. Abnormalities of fertilization due to egg ageing (chiefly polyspermy) are discussed in Chapter 6, and developmental anomalies from this cause in Book 2, Chapter 5 and Book 4, Chapter 7.

**Anomalies**

Eggs can show various abnormalities, attributable to errors in their development. Among the more common distinctive forms are 'giant' eggs, binucleate eggs, and ovulated primary oocytes (in animals that normally ovulate secondary oocytes). Giant eggs seem to have roughly twice the volume of normal eggs and so might possibly have arisen through failure of cytoplasmic division of oogonia; consistently, some giant eggs are also

Table 3.2. *Proportions of normal and abnormal pregnancies and litter sizes after insemination at various times in the rat*

| | Pregnancies (per cent) | Litter size | Abnormal pregnancies (per cent) |
|---|---|---|---|
| 10–12 hr before ovulation | 50 | 3.5 | 0 |
| Near time of ovulation | 83 | 6.7 | 11 |
| 6 hr after ovulation | 47 | 4.6 | 48 |
| 10 hr after ovulation | 22 | 1.8 | 79 |
| 12 hr after ovulation | 4 | 0 | 100 |

*Source*: Data from R. J. Blandau and E. S. Jordan, *Amer. J. Anat.* **68**, 275 (1941); A. L. Soderwell and R. J. Blandau, *J. Exp. Zool.* **88**, 55 (1941).

binucleate. Mononucleate giant eggs could have arisen through failure of both cytoplasmic and nuclear division of oogonia, in which case they would have twice the normal chromosome complement, but this has not yet been investigated. Binucleate eggs of normal size (as secondary oocytes or ootids) could be the result of failure of polar body emission (first or second, respectively); as primary oocytes they seem more difficult to explain, but could still have arisen from irregularities of oogonial divisions, since we have no information on the precise outcome of such events. Ovulated primary oocytes may have come from immature follicles ruptured passively by the ovulation of a nearby mature follicle; often they are smaller than normal which tends to support the idea.

We do not know whether these anomalous eggs can undergo fertilization and give rise to embryos. If they can, then both giant and binucleate eggs would be likely to develop as triploid embryos. Ovulated primary oocytes can be safely assumed unfertilizable (though not necessarily impenetrable to spermatozoa) except in the dog and fox.

Important non-distinctive anomalies that would require karyotyping for their detection include eggs with one extra chromosome. These, with fertilization and embryonic development, could be the origin of trisomies such as Down's Syndrome or mongolism. How the chromosome comes to be duplicated is still open to speculation, but the available evidence indicates that this occurs very early, when oogonia are entering meiosis and thus becoming oocytes, and it seems to happen in the later oogonia to make this change. These tardy oogonia, developing in their due turn later in reproductive life, account for the increased risks of mongolism in the over-35s.

An even more cryptic anomaly is seen in secondary oocytes (rabbit, mouse) that have been removed from the follicle as primary oocytes and have gone through the first meiotic division *in vitro*. These eggs look normal and can be fertilized *in vitro* or after return to a mated recipient animal. But either the fertilization is not normal, the male pronucleus failing to develop, or else subsequent development is very limited, since no offspring result. We can still only speculate on the reasons for these things.

## Parthenogenesis

The term parthenogenesis is used to distinguish the production of young without prior sexual intercourse and therefore without fertilization. (The definition does not fit gynogenesis in some fish, but that is another story – see Chapter 6.) We are still uncertain whether parthenogenesis ever occurs naturally in any mammal, or even could be induced by experimental intervention, though it is a normal process in several non-mammalian phyla, and is especially common among the insects. By contrast, the first few steps along the road – 'activation' of the egg (normally a consequence of sperm penetration) and cleavage of the egg in a manner closely similar to development after fertilization – have often been encountered in mammals, especially in rodents (Fig. 3.8). The course of events is identified as rudimentary parthenogenesis, because it is not known to proceed far, commonly for not more than a few cleavage divisions. Rudimentary parthenogenesis can easily be induced artificially and is of great interest to investigators because of the high degree of genetic homogeneity of the embryo, and the possibility of comparing haploid and diploid development in genetically almost identical material.

By the use of selected methods of activation induction and embryo culture, Matthew Kaufman and his colleagues have been able to observe apparently normal fetal development right up to the forelimb-bud stage in the mouse. It is difficult to know why it has not yet gone further. The notion that unopposed lethal genes would load the dice heavily against the continued progress of the parthenogenone is not as impelling as it used to be and alternative explanations are sought. 'Something is missing' is the consensus now and a number of suggestions have been made. Perhaps we may look forward to an authentic virgin birth in the not-too-distant future. (Further comments on parthenogenesis appear in Book 2, Chapter 1.)

In terms of their contribution to the gene complement of the embryo, eggs and spermatozoa are equal, but in several other ways the egg has the

Fig. 3.8. A scheme to show five possible routes of parthenogenetic development that may be taken by the secondary oocyte (at A) after emission of the first polar body: A → B, with emission of the second polar body, leading to haploid development; A → C, with 'immediate cleavage' (one daughter cell representing the second polar body), the result being again haploid development; A → D, suppression of emission of second polar body after completion of the second meiotic division, giving again haploid development; A → E, with suppression of the second meiotic division and delay of the first cleavage division, producing a diploid embryo; A → F, suppression of the second meiotic division, followed by normal cleavage, again opening the way to diploid development. (From M. H. Kaufman (1981) – see Suggested Further Reading.)

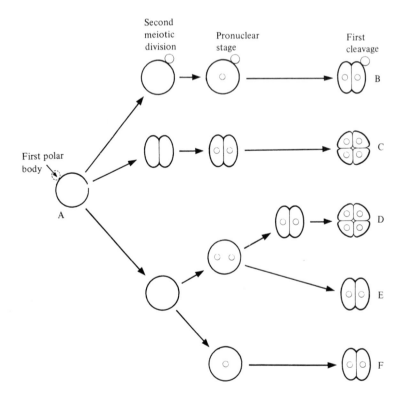

right to the centre of the stage. If needs must, the egg can provide all that is required for extensive development – perhaps even for full development – and no one has yet succeeded in getting a sperm nucleus to cleave in culture (though they have tried). The egg has been adopted by a remarkably high proportion of plants and animals. Not surprisingly the frontispiece to William Harvey's famous book on generation bears the motto *Ex ovo omnia* (Fig. 3.9).

### Suggested further reading

Ageing and reproduction: post-ovulatory deterioration of the egg. C. R. Austin. *Journal of Reproduction and Fertility*, Supplement **12**, 39–53 (1970).

Structure and function of the zona pellucida: identification and characterization of the proteins of the mouse oocyte's zona pellucida. J. D. Bleil and P. M. Wassarman. *Developmental Biology* **76**, 185–202 (1980).

Isolation, physicochemical properties, and macromolecular composition of zona pellucida from porcine oocytes. B. S. Dunbar, N. J. Wardrip and J. L. Hedrick. *Biochemistry* **19**, 356–65 (1980).

Parthenogenesis: a system facilitating understanding of factors that influence early mammalian development. M. H. Kaufman. *Progress in Anatomy* **1**, 1–57 (1981).

Fig. 3.9. Detail from the frontispiece of William Harvey's book *De Generatione Animalium*, published in 1651. The hands are those of Jove.

Normal postimplantation development of mouse parthenogenetic embryos to the forelimb bud stage. M. H. Kaufman, S. C. Barton and M. A. H. Surani. *Nature, London*, **265**, 53–55 (1977).

Parthenogenesis. U. Mittwoch. *Journal of Medical Genetics* **15**, 165–181 (1978).

Development of the cortical granules and the cortical reaction in rat and hamster eggs. D. Szollosi. *The Anatomical Record* **159**, 431–46 (1967).

Absence of centrioles in the first and second meiotic spindles of mouse oocytes. D. Szollosi, P. Calarco and R. P. Donahue. *Journal of Cell Science* **11**, 521–41 (1972).

*The Mammalian Egg*. C. R. Austin. Blackwell Scientific Publications; Oxford (1961).

Fine structure of mammalian oocytes and ova. A. T. Hertig and B. R. Barton. In *Handbook of Physiology – Endocrinology* II, Part 1, pp. 317–48. Ed. R. O. Greep. American Physiological Society; Washington, DC (1973).

# 4

# Spermatogenesis and spermatozoa

*B. P. SETCHELL*

Spermatozoa are produced in the testes by a process known as spermato-genesis. The first spermatozoa are released at puberty, but these represent the culmination of events that begin early in fetal life, and which are described in Chapter 1 of this Book. In this chapter, we will discuss the spermatogenic function of the testis from birth to old age, but concentrate on the mature adult, in particular during the time of active spermatogenesis in those species that show seasonal patterns of reproduction. Seasonal variations are discussed in Chapter 4 of Book 4 and the endocrinological function of the testis and the effects of testicular hormones on other organs of the body are covered in Chapter 3 of Book 2, as well as in Chapter 4 of Book 7 of the First Edition. What happens to the spermatozoa after they leave the testis, including their maturation in the epididymis and capacitation in the female reproductive tract, are dealt with in the next two chapters.

## The spermatozoon

### Structure

Spermatozoa were first described about 300 years ago by Leeuwenhoek in semen and in the testis. He noticed that each of the animalcules, as he called them, has a head and a motile tail. There was debate for a long time about the significance of the spermatozoa but people finally realized about 1830 that they were the part of the semen essential for fertility. Despite great advances made in the design of the optical microscope, little further could be discovered about the fine structure of spermatozoa until the introduction of the electron microscope, when the tail in particular was found to possess an exceedingly complex internal organization (Fig. 4.1). The spermatozoa of different species vary widely in size; those of humans, rabbits and the common domestic mammals are about 50 $\mu$m long while rodent spermatozoa are much longer, between 150 and 250 $\mu$m. The longest mammalian spermatozoa appear to be those of a tiny marsupial, the honey possum *Tarsipes rostratus*; those of some other marsupials are similar in size to rodent spermatozoa or to human spermatozoa (Fig. 4.2).

### Sperm head

The shape of the sperm head is characteristic of the species, hook-shaped in rats and mice but flattened and rounded in man and the domestic animals. The heads of human spermatozoa are about 5 $\mu$m long, 2.5 $\mu$m wide and 1.5 $\mu$m thick. In all mammals the sperm head consists of a highly condensed mass of DNA–protein called chromatin, in some way combined with strongly basic small proteins or peptides (mol. wt about 8000) known as protamines. The chromatin is stabilized by the formation of powerful disulphide bonds, which normally persist until fertilization but which can be broken experimentally with the assistance of chemical reagents. About the anterior half, or slightly more, of the head is covered by the acrosome (Fig. 4.3), which is a membrane-enclosed sac of enzymes important in the penetration of the egg by the spermatozoon. The exact shape of the

Fig. 4.1. The internal structure of a typical mammalian spermatozoon with the cell membrane removed. (From D. W. Fawcett. *Devel. Biol.* **44**, 394 (1975).)

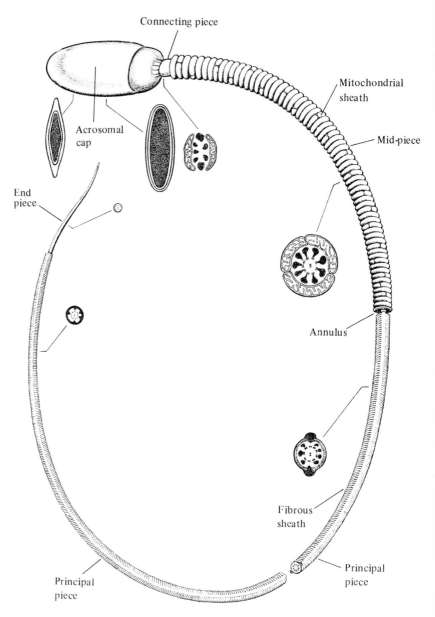

Fig. 4.2. A diagram of the spermatozoa of several marsupial and eutherian mammals. The acrosome and mid-piece are shaded. Spermatozoa 1–6 are those of Australian marsupials, while 7–13 are those of eutherian mammals. 1. honey possum *Tarsipes rostratus*; 2. marsupial 'rat' *Dasyuroides byrnei*; 3. short-nosed brindled bandicoot *Isoodon macrourus*; 4. tammar wallaby *Macropus eugenii*; 5. brush-tailed possum *Trichosurus vulpecula*; 6. koala *Phascolarctos cinereus*; 7. hippopotamus *Hippopotamus amphibius*; 8. man *Homo sapiens*; 9. rabbit *Oryctolagus cuniculus*; 10. ram *Ovis aries*; 11. golden hamster *Mesocricetus auratus*; 12. laboratory rat *Rattus norvegicus*; 13. Chinese hamster *Cricetulus griseus*. (Courtesy of L. M. Cummins.)

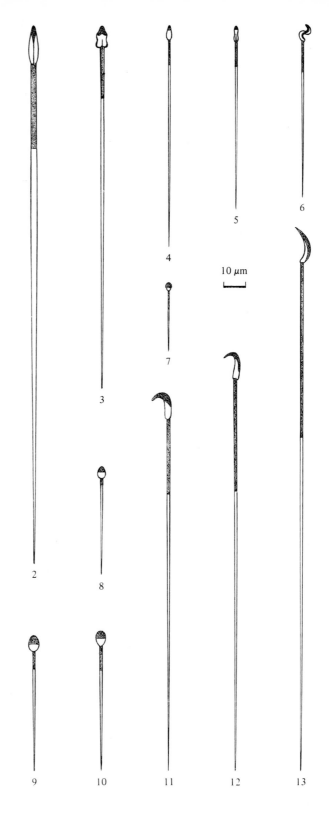

acrosome is also species-dependent, attaining really extreme forms in some rodents, but it is usually moulded over the nucleus, so that the acrosomal membrane can be said to have inner and outer parts. Like other cells, the whole spermatozoon including the acrosome is enclosed by a plasma membrane; as a cell it has a remarkably small amount of cytoplasm.

Behind the acrosome, there is a postacrosomal region which is important because it is in this region that the spermatozoon attaches and fuses to the egg (see Chapter 6). The plasma membrane in this region is underlain by a thick dense layer called the postacrosomal dense lamina or postacrosomal sheath. A narrow clear space between this sheath and the nucleus extends back to the posterior ring, where the plasma membrane is fused to the underlying nuclear envelope. The portion of the nuclear envelope beneath the acrosome and the postacrosomal sheath has no nuclear pores; these

Fig. 4.3. A sagittal section through the head of a primate spermatozoon. (From D. W. Fawcett, in *Frontiers in Reproduction and Fertility Control*. Ed. R. O. Greep and M. A. Koblinsky. Massachusetts Institute of Technology; Cambridge, Mass. (1977).)

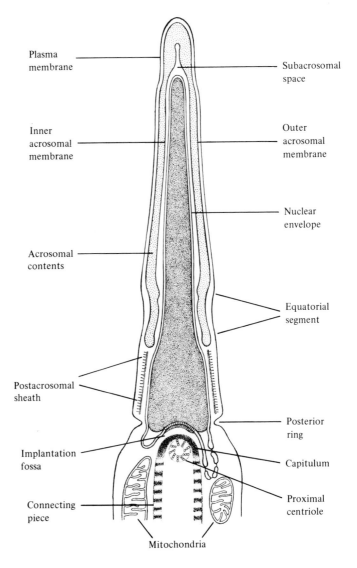

Plasma membrane

Subacrosomal space

Inner acrosomal membrane

Outer acrosomal membrane

Nuclear envelope

Acrosomal contents

Equatorial segment

Postacrosomal sheath

Posterior ring

Implantation fossa

Capitulum

Connecting piece

Proximal centriole

Mitochondria

are only found in the part of the nucleus that extends back into the neck region. The part of the head nearest to the mid-piece is highly specialized to form an implantation fossa, in which pores are also absent. This region of the nuclear membrane is covered on its outer surface by a thick layer of very dense material which constitutes the basal plate.

*Sperm tail*

The tail or flagellum is usually divided into a mid-piece, principal piece and end-piece. The mid-piece extends from the head to as far as the end of the mitochondrial helix. The portion nearest to the head is called the connecting piece, and it joins the motile apparatus to the nucleus. It is an exceedingly complex structure. At its proximal end, the dense, convex joint-like or articular region called the capitulum is attached to the basal plate of the nucleus by a series of fine filaments. These filaments are broken or dissolved by agents or events that separate the head from the tail. Extending back from the capitulum are nine distinctively segmented columns, which at their further ends overlap the tapering ends of the nine dense outer fibres of the flagellum, to which they are firmly united. The neck region is usually quite slender and there are no organelles there, except for a fold or scroll of nuclear envelope and one or two longitudinally orientated mitochondria. The rest of the mitochondria are lined up end-to-end to form a tight helix around the dense outer fibres (Fig. 4.4).

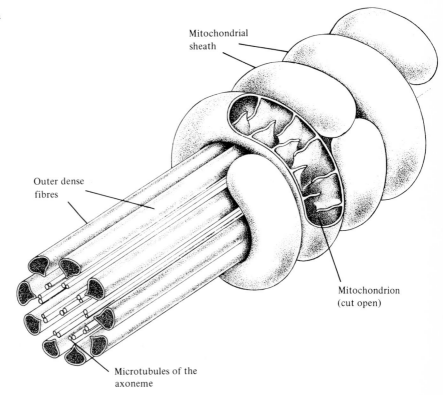

Fig. 4.4. The mid-piece of a eutherian spermatozoon with the cell membrane removed. (From D. W. Fawcett, as in Fig. 4.3.)

Mitochondrial sheath

Outer dense fibres

Mitochondrion (cut open)

Microtubules of the axoneme

There is wide variation between species in the number of gyres in the mitochondrial helix, ranging from as few as 15 in man to more than 300 in some rodents. Otherwise the mitochondria in most species appear like those from other cells.

At the centre of the tail is the axoneme or axial filament complex. It consists of two central microtubules surrounded by a cylinder of nine evenly spaced doublet microtubules. Each doublet microtubule consists of two subunits: subfibres A and B. Subfibre A has attached to it two lines of dynein arms projecting towards the next doublet (Fig. 4.5); these arms, which have strong ATPase activity, are probably the basis of sperm motility and if they are absent the spermatozoa are immotile. There are also two slender nexin links joining each doublet to the adjacent one. Radial spokes join each doublet to a helical sheath around the central pair of microtubules. The microtubules are formed mostly of a protein, called tubulin, which is identical to the protein found in the chromosomal spindle of dividing cells. Between the axoneme and the mitochondrial helix are the nine dense outer fibres. Each is continuous anteriorly with one of the columns of the connecting piece and runs just outside its corresponding axonemal doublet. The nine dense outer fibres differ from one another in their structure, both in size and cross-sectional shapes, and there is also

Fig. 4.5. A schematic reconstruction of the axoneme of a eutherian spermatozoon. There is no evidence that the dynein arms are rectangular, but they are discontinuous projections. (From D. W. Fawcett, as in Fig. 4.3.)

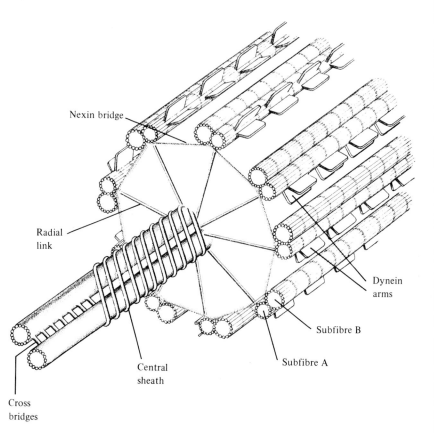

Nexin bridge

Radial link

Cross bridges

Central sheath

Subfibre A

Subfibre B

Dynein arms

considerable variation between species in their thickness. Near their end, the outer fibres appear to be fixed to the wall of the corresponding doublet. There is no ATPase activity in the outer fibres which appear to be formed from several polypeptides of various sizes.

In the principal piece, beyond the mitochondrial helix, the axonemal filaments and the dense outer fibres continue, surrounded by a fibrous sheath (Fig. 4.6). This consists of a number of circumferentially oriented ribs running between two longitudinal columns, which extend along the entire length at either side of the tail. In its initial portion the sheath is fixed to two of the dense outer fibres (numbers 3 and 8), which terminate well before the rest of the fibres. As the sperm tail tapers, the columns become smaller and the ribs thinner; the fibrous sheath ends several microns from the tip. The fibrous sheath is made up of a single protein with a mol. wt of about 80 000.

The end-piece, beyond the end of the fibrous sheath, consists of the 9 + 2 microtubules, covered only by the plasma membrane.

## Metabolism

Spermatozoa are able to utilize a wide range of substrates, both aerobically and anaerobically. Simple sugars such as glucose, fructose or mannose can

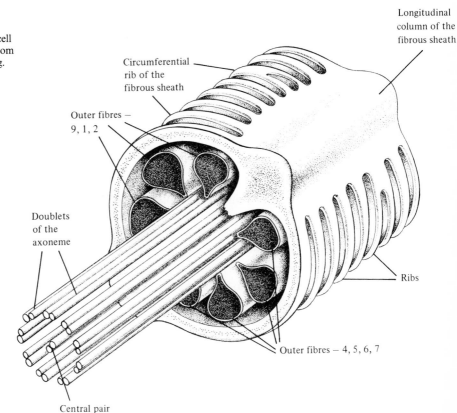

Fig. 4.6. The principal piece of a eutherian spermatozoon with the cell membrane removed. (From D. W. Fawcett, as in Fig. 4.3.)

Longitudinal column of the fibrous sheath

Circumferential rib of the fibrous sheath

Outer fibres – 9, 1, 2

Doublets of the axoneme

Ribs

Outer fibres – 4, 5, 6, 7

Central pair

be broken down by glycolysis to lactic acid, or oxidized to carbon dioxide and water, depending on the circumstances. However the spermatozoon is unusual in that the rate of glycolysis is not less under aerobic conditions, in other words, they do not show a Pasteur effect. Spermatozoa can also oxidize many other simple organic substances, such as lactate, pyruvate, volatile fatty acids and citric acid cycle intermediates such as succinate. There was some evidence that human spermatozoa had a very poor oxidative capacity and could only maintain motility in the presence of a glycolysable substrate, but more recent experiments have not substantiated this idea. Instead it would appear that human spermatozoa are particularly sensitive to the Tris buffer used in the earlier experiments.

The immediate source of energy for motility appears to be the pool of adenine nucleotides: ATP, ADP and AMP. The spermatozoon also contains high concentrations of carnitine, which in its acetylated form probably serves as an important metabolic reserve. However, it is probably more accurate to say that motility controls metabolism, rather than the other way round. Apart from motility, sperm metabolism does not appear to be under any particular control, and it is unlikely that the spermatozoon finds itself without adequate supplies of exogenous substrate at any point in its journey in male and female genital tracts.

Ejaculated spermatozoa have very little synthetic capacity, but the same is not true at the time when they leave the testis, as several important metabolic changes take place during passage through the epididymis. If ejaculated spermatozoa are suddenly subjected to low temperature they undergo a reaction known as 'cold shock', involving irreversible loss of viability and increased permeability of the cell membrane. This is probably an effect of temperature on the lipids of the sperm membrane, rather than a specific metabolic lesion. Testicular spermatozoa are not susceptible to cold shock.

*Motility*

The flagellar protein that contributes most to the conversion of chemical energy to mechanical movement is dynein, which has a mol. wt of about 500000 and exhibits ATPase activity; it is located in the arms of the doublet microtubules. The microtubules themselves are not contractile; in fact they are incapable of shortening. Instead, adjacent doublets slide past one another, propelled by changes in conformation of the dynein arms. The dynein arms on subfibre A have two different positions: extended and flattened. ATP is bound to the dynein arm in the flattened position; the arm begins to extend from the subfibre as the ATP is broken down and binds to the tubulin lattice of subfibre B of the adjacent microfilament. The arm then changes its angle of attachment to adopt the 'rigor' position, as the products of ATP hydrolysis are released. Production or addition of new ATP causes the dynein arm to detach and return to the flattened position (Fig. 4.7).

The movements of the microtubules of one half of the axoneme are in synchrony, but they are out of phase with those of the microtubules in the other half, so that this sliding movement is converted into a bending motion, although little is known about how the sliding movements between tubules are coordinated to produce a propagated wave of bending motions down the tail. Some sort of interaction between the central pair of microtubules and the radial spokes may be the basis for a regulatory mechanism.

## Testes

In all mammals and birds, there are two ovoid testes, of approximately equal size, one on each side, but in many lower animals there are multiple testes and in a few species a single testis. The weight of the two testes in mammals ranges from a lower value of about 0.02 per cent of the body weight in the gorilla to an upper value of about 4 per cent of body weight in the honey possum which as we have already mentioned also has the

Fig. 4.7. A diagram of the postulated cycle of form changes that the dynein arms execute in association with movement of micro-filaments. (From P. Satir, in *The Spermatozoon*. Ed. D. W. Fawcett and J. M. Bedford. Urban and Schwarzenberg; Baltimore (1979).)

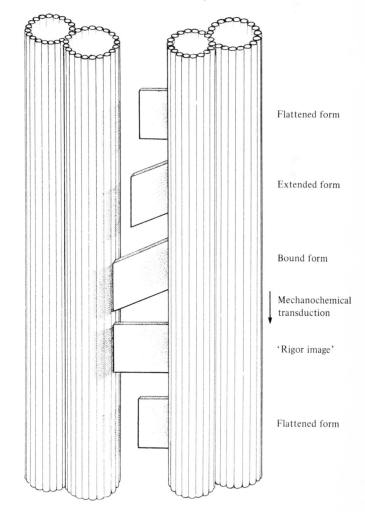

Flattened form

Extended form

Bound form

Mechanochemical transduction

'Rigor image'

Flattened form

longest spermatozoa. Humans have quite small testes, about 0.08 per cent of body weight, while in most common domestic animals and laboratory animals the two testes comprise between 0.5 and 1.5 per cent of body weight.

### Capsule

The parenchyma of the mammalian testis is encased in a tough fibrous capsule, which in some species contains myoid or smooth-muscle-like elements. The capsule in these animals is contractile and may be important in maintaining the appropriate pressure within the tissue. Most of the larger arteries and veins running to the testis are also found in the capsule.

### Seminiferous tubules

If the capsule is removed we can see that the testis is made up largely of closely packed seminiferous tubules. Each tubule is between 200 and 250 $\mu$m in diameter in most species, but in some Australian marsupials, the tubular diameter is about 450 $\mu$m. The exact proportion of tubular tissue can vary from about 90 per cent in rats, rams and bulls to about 60 per cent in pigs, horses and some marsupials. In many species, including the human, the tubules are arranged in lobules separated by bands of fibrous tissue, but in others, for example most rodents, there are no

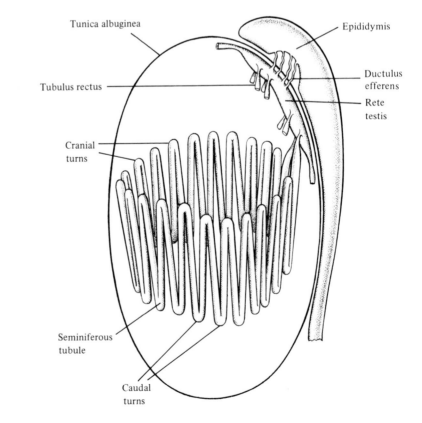

Fig. 4.8. Diagram to show the arrangement of one seminiferous tubule in a rat testis, and its relation to the rete testis. As there are about 30 tubules in the species, the rete would actually have about 60 tubuli recti opening into it. (From Y. Clermont and C. Huckins. *Am. J. Anat.* **108**, 79 (1961).)

Tunica albuginea

Tubulus rectus

Cranial turns

Seminiferous tubule

Caudal turns

Epididymis

Ductulus efferens

Rete testis

subdivisions, the 30-odd tubules in the rat, being tightly coiled, lie apparently at random throughout the organ. Each tubule has the form of a loop, both ends of which open into the rete testis (Fig. 4.8). Very occasionally, branched or blind-ended tubules can be found.

### Interstitial tissue

The seminiferous tubules are not penetrated by any blood vessels or nerves for these are confined to the spaces between the tubules; here they form part of the interstitial tissue, which also contains the Leydig cells, the cells responsible for almost all the de-novo synthesis of steroids in the testis (see Chaper 4, Book 3, and also Chapter 4, Book 7 in the First Edition). In

Fig. 4.9. A diagram of the structure of the interstitial tissue of three types of testes. Type I is found in rodents; in some species, e.g. the rat, the groups of Leydig cells around the capillaries are not surrounded by a continuous endothelial layer, in others, e.g. the guinea pig, which is not shown here, the endothelial layer is continuous. Type II is found in human and sheep, and Type III in pig and some other species. (From D. W. Fawcett. *Adv. Biosci.* **10**, 83 (1973).)

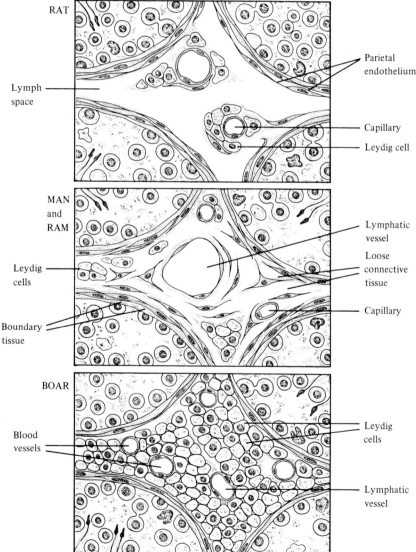

the interstitial tissue one also sees a large number of macrophages and a complex network of lymph vessels, which are large sinusoidal spaces in rodents but discrete ducts in men, rams and bulls. In boars, in which the Leydig cells form a tightly-packed mass, the lymphatic vessels are small and insignificant (Fig. 4.9), but it is interesting that the rate of lymph flow per unit weight of testis is highest in pigs and lowest in rats, with sheep intermediate.

### Rete testis

Both ends of every seminiferous tubule open into the rete testis, which is an intercommunicating system of channels lined with a flattened epithelium, situated centrally in the testis of animals such as rams, bulls and boars, and at one side, near the epididymis, in men and rodents. The ends of the seminiferous tubules are modified into straight tubules or tubuli recti in some species, and are lined with rete-like epithelium, but in others there is a transitional zone, lined with cells like Sertoli cells, opening directly into the rete.

The fluid from the seminiferous tubules is modified in composition in the rete (see p. 81), probably because of the greater permeability of its epithelium, and some forms of damage to the tubules, especially immuno-logical injury, often begin at the rete and then spread into and along the tubules.

### Efferent ducts and epididymis

The efferent ducts, or ductuli efferentes, usually three to eight in number, lead from the rete testis into the epididymis, where they unite to form the epididymal duct, a single much convoluted tube that eventually gives rise to the ductus deferens or vas deferens. The epithelium of the efferent ducts and the first part of the epididymis is strongly fluid-resorptive; the cells lining the middle part of the epididymal duct or corpus appear to have secretory activity, whereas in the tail, they are low and appear much less active. (More attention is given to these duct systems in Chapter 5.)

### The scrotum and descent of the testis

In many mammals, but by no means all members of this class, the testes descend into a scrotum during fetal or early post-natal life (see Fig. 4.10). This does not happen in any other animals and the reason for the migration is obscure.

### Temperature of the testis

Whatever the explanation may be for the descent of the testis, once outside the abdominal cavity, the organ finds itself in a cooler environment; it then becomes susceptible to damage if its temperature is raised to that of the body cavity (see p. 99). Whether this temperature dependence is a cause or consequence of testicular descent is not known. The scrotal testis is between 4 and 7 °C cooler than the general deep body temperature, and the scrotal skin in many species is well furnished with sweat glands to keep the temperature down by evaporative heat loss, even in those animals such as the sheep in which sweating is not an important mechanism in general thermoregulation.

### Spermatic cord

As the testis descends, the testicular artery becomes elongated, and in mammals with pendulous scrota, such as sheep and cattle, the artery outside the abdominal cavity is further lengthened and coiled, so as to form a vascular cone in the spermatic cord. In this cone there can be up to 7 m of artery coiled into a structure only about 10 cm long. The coiled artery is surrounded by the pampiniform plexus, a network of multiple small veins arising from veins leading out of the testis. The veins of the plexus reunite to form usually a single vein near the inguinal canal. In marsupials,

Fig. 4.10. A diagram showing the position of the testes in some mammals. 1. Just behind kidney: elephants, hyraxes. 2. At posterior end of abdominal cavity: dolphins, whales, armadillos, sloths and anteaters. 3. Against ventral abdominal wall: hedgehog and some other insectivores. 4. At base of tail: mole, shrew and some bats. 5. In non-pendulous scrotum underneath anus: pigs, horses, rats, rabbits, dogs, and cats. 6. In pendulous scrotum: sheep, cattle and other ruminants, some primates. (From F. N. Carrick and B. P. Setchell, in *Reproduction and Evolution*, ed. J. H. Calaby and C. H. Tyndale-Biscoe. Australian Academy of Science; Canberra. (1977).)

the artery, instead of being a single elongated vessel, divides to form a plexus of up to 200 parallel branches, which reunite before dividing again to supply the testis. The arterial branches in the plexus are interspersed among a similar number of branches of the testicular veins.

This complex vascular system constitutes a highly efficient countercurrent heat exchanger in which the arterial blood is precooled before it reaches the testis to the temperature of the scrotal skin, and the venous blood is warmed to body temperature before return to the abdomen (Fig. 4.11). Similar mechanisms exist in the fins of whales and seals, and in the legs

Fig. 4.11. The temperature, mean blood pressure and pulse pressure, and the concentration of testos- terone (T), in the blood vessels leading towards and away from the testis of a ram. (From B. P. Setchell, in *Reproduction in Domestic Animals*, ed. H. H. Cole and P. T. Cupps. Academic Press; New York and London. (1977).)

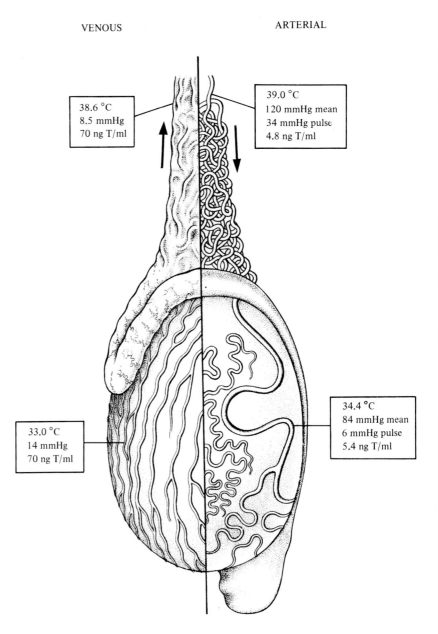

VENOUS

ARTERIAL

38.6 °C
8.5 mmHg
70 ng T/ml

39.0 °C
120 mmHg mean
34 mmHg pulse
4.8 ng T/ml

33.0 °C
14 mmHg
70 ng T/ml

34.4 °C
84 mmHg mean
6 mmHg pulse
5.4 ng T/ml

of Arctic birds, but in these the function is to prevent excessive heat loss through the extremities under cold conditions. The primary purpose of the spermatic cord is not known, but the structure does ensure that the whole testis is kept at a lower temperature than organs within the body cavity. However, a spermatic cord is present, although it is not so well developed, even in mammals such as the hedgehog and armadillo in which the testis migrates but stays within the abdomen and therefore remains at deep body temperature. Therefore it seems unlikely that the primary reason for this structure lies in thermoregulation. In animals like elephants and hyraxes, in which the testis does not migrate at all from its embryological starting-point, the artery and vein remain single, straight and separate.

Another effect of the spermatic cord in mammals with scrotal testes is to eliminate the pulse pressure from the arterial blood going to the testis (see Fig. 4.11), but the significance of this is even less clear. It has also been suggested that the close apposition of veins and artery would allow exchange of diffusible substances from one to the other, enabling a high local concentration of some important ion, metabolite or hormone to be built up within the testis. Unfortunately, there is no experimental evidence that this intriguing mechanism operates for testosterone or any other known product of the testis, although radioactive water and some other substances have been found to cross to some extent from testicular vein to artery. (In the female, the homologue of the pampiniform plexus allows uterine prostaglandins to pass from the utero-ovarian veins into the ovarian artery, thus bringing about local luteal regression – see Book 3, Chapter 5.)

An alternative explanation for the descent of the testis is that the organ just follows the epididymis into the scrotum because the tail of the epididymis needs to be in a cool place for sperm maturation and storage. Again this challenging idea has not been supported by experimental evidence. When the epididymis was returned to the abdomen, but the testis left in the scrotum, the animals remained fertile for a long time, even though some regressive changes in epididymal function could be demonstrated.

## Seminiferous tubules and spermatogenesis

In the seminiferous tubules, there are two types of somatic cells: the myoid or smooth muscle-like cells and the Sertoli cells, and five types of germ cells: spermatogonia, primary and secondary spermatocytes, spermatids and spermatozoa.

### Myoid cells

These cells constitute an important part of the boundary tissue or wall of the seminiferous tubules along with non-cellular material, most of which lies between the myoid cells and the other tubule cells as a sort of basement membrane. Myoid cells are probably responsible for the peristalsis-like movements of the tubules, but may also exert an important stimulatory influence on the Sertoli cells, as suggested from the interaction between these two cell types in culture.

### Sertoli cells and the blood–testis barrier

The Sertoli cells (Fig. 4.12) occupy a pivotal position in the seminiferous epithelium in that they are the only cells to extend from the outer wall of the tubule to the lumen, and to have contacts with all the other cell types in the epithelium. They have bases that rest against the boundary tissue. The spermatogonia lie between the Sertoli cells and the boundary tissue, and the other germ cells are either sandwiched between adjacent pairs of Sertoli cells or actually embedded in crypts in the luminal surface of the Sertoli cell cytoplasm, which as a result has a very extensive and highly irregular outline. The Sertoli cells have been likened to trees in an orchard, because of their regular spacing when the tubules are viewed from the side, with the other cells corresponding to the fruit, but this analogy should not be taken too far.

Fig. 4.12. A diagram of the fine structure of a Sertoli cell and part of another, showing how they extend from the boundary tissue (bottom) to the lumen (top). They enclose the germ cells in the basal compartment between themselves and the boundary tissue, in different niches of the adluminal compartment between pairs of Sertoli cells and in crypts in their luminal surfaces. (From D. W. Fawcett, in *Handbook of Physiology, Section 7 Endocrinology, vol. 5: Male Reproductive System.* Ed. D. W. Hamilton and R. O. Greep. American Physiological Society; Washington. (1975).)

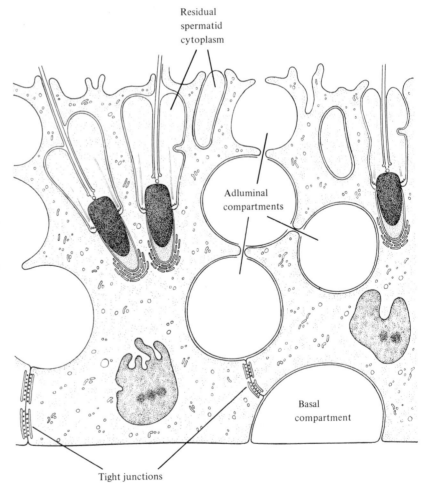

Residual spermatid cytoplasm

Adluminal compartments

Basal compartment

Tight junctions

The Sertoli cells are diploid, with characteristically indented nuclei, and prominent nucleoli. They apparently do not divide after puberty, although just before then there seems to be a burst of mitotic activity. They have numerous mitochondria, and a complex Golgi apparatus. Both rough and smooth endoplasmic reticulum are present, in most species predominantly the smooth type, and some specialized aggregations of this lie immediately adjacent to the developing acrosome of the spermatid. There are often abundant lipid droplets, but the number varies widely among species, and in the rat in which they are common, they vary with the stage of the seminiferous cycle, being most obvious immediately after sperm release. Human Sertoli cells also contain two different types of crystalline deposits, Charcot-Böttcher and Spangaro crystals of unknown significance.

*Junctional structures.* The Sertoli cells form a remarkable range of specialized contacts with the various germ cells and with each other. There are desmosome-like junctions which first appear between the Sertoli cells and B-spermatogonia but are most extensive and highly developed between the Sertoli cells and large pachytene spermatocytes. They gradually disappear so that fewer are seen during meiosis and they are rare once the germ cells become spermatids. These junctions may be important in the

Fig. 4.13. The tubulo-bulbar complex of a rat spermatid. (From L. Russell and Y. Clermont. *Anat. Rec.* **185**, 259 (1976).)

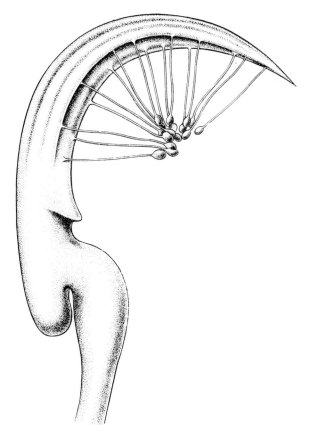

translocation of the germ cells towards the lumen of the tubule. Within the desmosomes, there are gap junctions to be found which disappear as the germ cells pass through the meiotic divisions, and which may be involved in direct cell-to-cell transfer of information or metabolites.

The surfaces of the Sertoli cells also show unique features, known as ectoplasmic or junctional specializations, which first appear opposite the pachytene spermatocytes, are less obvious at about the time of meiosis, and become apparent again as the spermatids begin to elongate. These specializations are most prominent next to the acrosome region of the maturing spermatids.

Near the time of sperm release, the ectoplasmic specializations lose their contacts with the cell surface and may be recycled, because it is shortly afterwards that a new set appear near the pachytene spermatocytes and increase in extent near the elongating spermatids. These structures may be involved in maintaining a tight relationship between the Sertoli cells and certain developing germ cells, or in stiffening the surface zone of the Sertoli cell where the maturing spermatids indent its luminal surface. As these specializations move away from the late spermatids, a fourth type of structure appears, the tubulo-bulbar complex (Fig. 4.13). Bristle-coated pits develop in the Sertoli cell, into which projections of spermatid cytoplasm extend and then form bulbous dilatations. Just prior to sperm release in the rat, the Sertoli cell withdraws from around the head of the spermatid, and new tubulo-bulbar complexes form along its convex edge. These are then resorbed and the spermatozoon is released.

Finally – and in the present state of knowledge, most significantly – pairs of Sertoli cells are also joined to one another by junctional complexes that have an extremely intricate morphology and evidently bring the two cell surfaces very close together, for they block the penetration of electron-opaque markers, such as horseradish peroxidase and lanthanum. The intercellular spaces between pairs of Sertoli cells are thus divided into a *basal compartment*, between the boundary tissue and the junctions, and an *adluminal compartment* above the junctions (Fig. 4.12). Spermatogonia as they rest on the boundary tissue exist in the basal compartment, while primary spermatocytes and later spermatogenic cells are in the adluminal compartment.

*Composition of tubule fluid.* The adluminal compartment presumably communicates reasonably freely with the fluid in the lumen of the tubules, and there is now abundant evidence that this fluid is quite different in composition from blood plasma, or indeed from lymph in the lymphatic vessels of the spermatic cord which is assumed to reflect the composition of the extracellular fluid of the interstitial tissue (Fig. 4.14). The tubular fluid is much higher in potassium, lower in sodium, and much lower in protein and glucose than blood plasma; it contains unusually high concentrations of inositol, glutamic acid and certain other amino acids, as well as a number of interesting peptides or proteins. These include a potent mitotic stimulator, an inhibitor of acrosin and, in some species but not all, androgen-binding protein (ABP).

ABP binds both testosterone and dihydrotestosterone, but its function in the testis is obscure. It certainly does *not* concentrate testosterone within the tubules, as has been suggested; the concentration of this steroid is no greater in tubular fluid than in extracellular fluid from the interstitial tissue. The most likely role of ABP is to transport androgen to the first part of the epididymis, where the protein–steroid complex is taken up from the lumen by the lining cells, the protein degraded and the steroid released. ABP is produced by the Sertoli cells, and appears to be a specific response to FSH stimulation in immature animals (see Book 7, Chapter 2).

Some characteristics of the tubular fluid change as it passes into the rete testis: in particular, the potassium concentration falls and the sodium rises, but neither reaches the level in blood plasma. There is no conclusive evidence for net secretion of fluid in the rete. The changes in composition are presumably due to a difference in permeability or secretory capacity between the cells lining the rete and the Sertoli cells which are believed to secrete the tubular fluid.

*Blood–testis barrier.* These differences in concentration between the fluids inside and outside the tubule would not be maintained if there was ready exchange across the tubule wall, and the existence of a blood–testis barrier in the tubules has been confirmed by studies on the rate of passage of a range of radioactively labelled markers from blood plasma into the fluid.

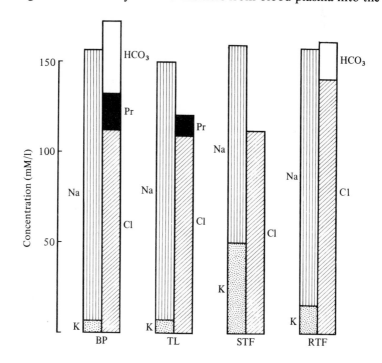

Fig. 4.14. The composition of blood plasma (BP), testicular lymph (TL), seminiferous tubule fluid (STF) and rete testis fluid (RTF) of rats. Cl, chloride; $HCO_3$, bicarbonate; K, potassium; Na, sodium; Pr, protein. (From B. P. Setchell. *The Mammalian Testis*. Elek; London. (1978).)

These studies show that, in general, lipid-soluble hydrophobic molecules penetrate more rapidly than hydrophilic substances (Fig. 4.15), and that there are probably specific carrier systems of the facilitated diffusion type for some substances, for example glucose and testosterone. The blood–testis barrier is in many ways analogous to the blood–brain barrier, but the latter differs in that it is located almost entirely in the capillary endothelial cells. The blood–testis barrier is established at or shortly before puberty, and is resistant to many treatments that break down permeability barriers elsewhere in the body.

The blood–testis barrier is important because it normally prevents proteins, including protein hormones (see p. 97) and antibodies from coming into contact with the germ cells from spermatocytes onwards, and, in the reverse direction, it prevents the proteins of the haploid spermatogenic cells from causing an immune reaction (see p. 96). The barrier also regulates the entry into the testis tubules of many small hydrophilic compounds, with important metabolic and toxicological consequences. However its primary function is probably to create inside the tubules the appropriate conditions for meiosis, although it must be admitted that we do not yet know what exactly these conditions are (but see Chapter 1 for some possibilities).

*Germ cells*

*Gonocytes.* When the migration of the primordial germ cells to the genital ridge is complete, in the male they take up their position in the centres of the sex cords, which become the seminiferous tubules; here they are surrounded by sustentacular or Sertoli cells and a prominent basement membrane. The male gonocytes do not then divide as their female counterparts do, but they remain inactive until shortly before puberty. At this time they migrate between the Sertoli cells to reach the boundary tissue of the tubule, and enter upon an indefinite period of mitotic multiplication.

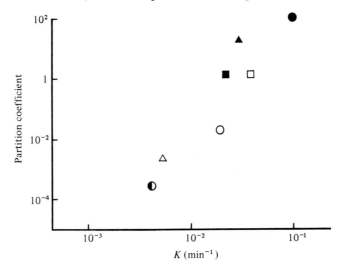

Fig. 4.15. The relationship between lipid solubility (measured as the partition coefficient between chloroform and phosphate buffer) and the entry rate ($K$) into rat rete testis fluid of several sulphonamides and barbiturates, all with mol. wts between 150 and 250. (Data from Okumura et al. *J. Pharmacol. Exptl. Therap.* **194**, 89 (1975).)

*Spermatogonia.* Once these divisions resume, the gonocytes are usually referred to as spermatogonia; in the adult, these are diploid cells, dividing mitotically, situated in the basal compartment of the tubule between the Sertoli cells and the boundary tissue. The spermatogonia are classified into three types: A, Intermediate and B. The several subdivisions of the A spermatogonia are indicated by subscripts 0, 1, 2, 3 or 4. Spermatogonial mitoses are of two types: the random divisions of the stem cells or $A_0$ spermatogonia (which are also called $A_s$ spermatogonia), and the synchronized divisions of $A_1$, $A_2$, $A_3$, $A_4$, Intermediate and B spermatogonia, which are coordinated with other events in the spermatogenic cycle (see p. 88). Subtle morphological differences can be discerned between the different types of spermatogonia, especially between A, Intermediate and B, but the various sub-types of A spermatogonia can usually be distinguished only by the stages of the cycle in which they are found. In the synchronized divisions, cytoplasmic links persist between the daughter cells after each division, with the result that clones of joined cells are formed.

Fig. 4.16. The two suggested schemes for the origin of spermatogonia and the relationship between the stem cells ($A_0$ or $A_s$ spermatogonia) and the spermatogonia ($A_1$, $A_2$, $A_3$, $A_4$, In and B) whose divisions are synchronized with other events in the spermatogenic cycle. In the scheme on the right, a stem cell divides to give a pair of spermatogonia ($A_{pr}$) and these divide to form four aligned spermatogonia ($A_{al}$) and this division may be repeated. PL, preleptotene spermatocyte.

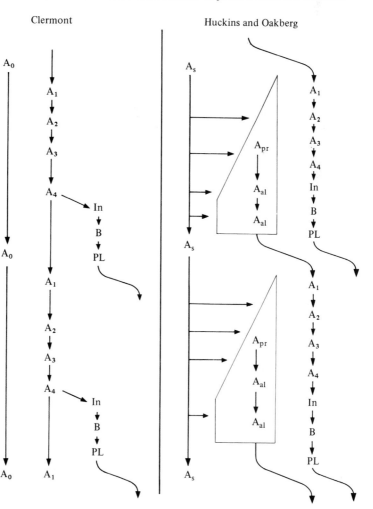

The origin of the $A_1$ spermatogonia is the subject of some dispute (Fig. 4.16). Yves Clermont believes that $A_0$ spermatogonia divide in the adult only when the other spermatogonia are destroyed, for example by X-irradiation, and that when the $A_4$ spermatogonia divide they give rise either to Intermediate spermatogonia or a new generation of $A_1$ spermatogonia. On the other hand, Claire Huckins and Eugene Oakberg believe that the $A_0$ or $A_s$ spermatogonia are continually dividing at a slow rate, uncoordinated with other events in the spermatogenic cycle. If these divisions give rise to a pair of separate daughter cells, they then remain as randomly dividing stem cells, but if their cytoplasmic connection is not broken, then at a fixed time in the spermatogenic cycle, all the pairs or sets of four or eight cells, which they go on to form, become $A_1$ spermatogonia. From then onwards, their divisions are synchronized with the other events in the spermatogenic cycle. There is no general agreement on which of these two schemes is the true one, but it is agreed that if damage does occur to the germinal epithelium, then the rate of division of the stem cells increases so that the tubules can be repopulated.

*Primary spermatocytes.* Immediately after the last mitotic division of the B spermatogonia, the daughter cells, still as a clone linked by cytoplasmic bridges, double their content of DNA as if they were about to divide again mitotically. However, instead of doing so they embark on the long meiotic prophase which leads eventually through two meiotic divisions to the production of four haploid spermatids from each spermatocyte. Just how do these cells know that the time has come for a change? We do not know, but it may be relevant that immediately after the DNA synthesis (during the preleptotene phase), the spermatocytes enter a part of the meiotic prophase known as leptotene and move away from the boundary tissue; the Sertoli cells on either side meet and form a second set of junctional complexes between the spermatocytes and the boundary tissue. For a while, the developing germ cells are in an intermediate compartment, then the junctions on the luminal side separate and the germ cells are safely through into the adluminal compartment to complete meiosis (Fig. 4.17).

In the next stage, zygotene, corresponding pairs of chromosomes come together and gather into a bouquet-like pattern. Distinctive linear synaptonemal complexes appear between the pairs of chromosomes. During the following stage, pachytene, the longest part of the meiotic prophase, each chromosome becomes thicker and begins to show a longitudinal division, except at the level of the centromeres. Finally in diplotene, when the nucleus reaches its maximal size, the duplication of the chromosomes is complete so that complete tetrads are formed. The X and Y chromosomes form a spherical heterochromatic body, quite distinct from the autosomes. Pachytene is a stage at which the spermatocytes are very susceptible to damage and widespread degeneration can occur at this stage.

*Meiosis and secondary spermatocytes.* Diakinesis, the separation of the homologous members of the paired doubled chromosomes, marks the end of the first meiotic division. It occurs quite rapidly and each primary spermatocyte gives rise to two secondary spermatocytes, each containing half the chromosomes of the parental cells, although by this time, each chromosome is already doubled. The life-span of the secondary spermatocytes is short; each cell soon divides again, with the two halves of each paired chromosome separating, to yield two haploid spermatids.

Fig. 4.17. Passage of the developing germ cells through the blood–testis barrier. The arrows indicate points of entry of electron opaque tracers, the presence of which in the spaces between the cells is signified by the regularly arranged dots. SN, Sertoli cell nucleus. (From M. Dym and J. C. Cavicchia. *Biol. Reprod.* **17**, 390 (1977).)

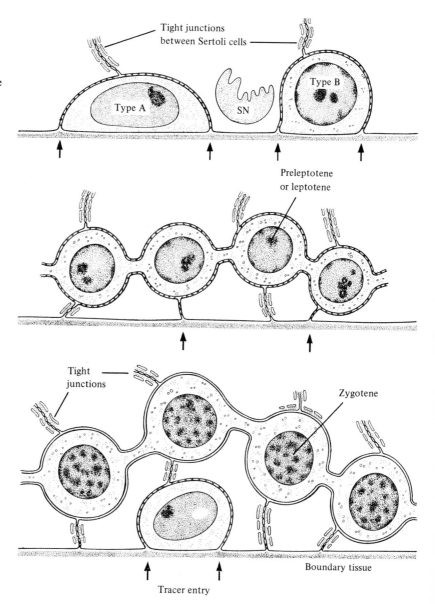

Fig. 4.18. Development of
a guinea pig spermatid.
(From D. W. Fawcett,
W. Anderson and D. M.
Phillips. *Devel. Biol.* **26**,
220 (1971).)

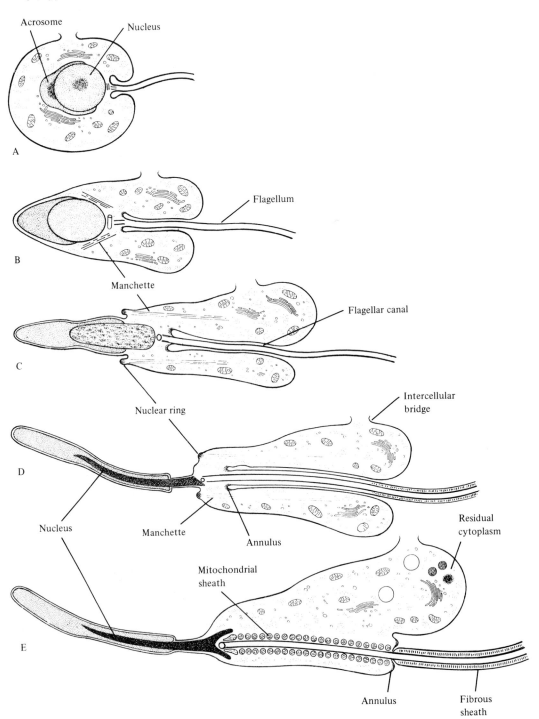

*Spermatids.* At the end of the second meiotic division, the young spermatids find themselves associated with an older generation of spermatids formed one cycle earlier, but still dependent on the Sertoli cells. The development of these two generations continues in synchrony, with the older ones embedded in crypts in the luminal surface of the Sertoli cell cytoplasm and the younger ones sandwiched between adjacent pairs of Sertoli cells, as they had been while they were spermatocytes and also during the meiotic divisions. (The development of the spermatid from meiosis to the detachment of the spermatozoa is known as spermiogenesis or spermateliosis). To begin with the young spermatids are rather ordinary-looking round cells with a plain nucleus, but quite soon things begin to happen (Fig. 4.18). The most obvious change is the condensation of the nuclear chromatin, so that the nucleus becomes denser and less reactive to Feulgen stain. In the human this process can often leave vacuoles which show up in electron micrographs. During condensation, the ultrastructure of the chromatin changes from a beaded pattern to a smooth fibre.

The Golgi apparatus of the spermatid gives rise to several cisternae containing granules. These cisternae fuse together to form a single acrosomal vesicle and granule. The granule and its enclosing vesicle then migrate towards the nucleus and become attached to it, while further material is added from the Golgi apparatus. The acrosome then spreads over part of the nuclear surface.

At the end of the meiotic divisions the two centrioles migrate to a position on the side of the nucleus opposite to the developing acrosome. One centriole assumes a radial alignment and the other positions itself at right angles to its fellow. The former then starts to grow the axoneme of the future sperm tail from its outer end, while the other gives rise to the neck or connecting piece which joins the tail to the nucleus. The dense outer fibres arise as lateral outgrowths from the walls of the corresponding doublets, and then separate from them, except at the far ends.

As the axoneme of the tail elongates, it protrudes from the main body of the cytoplasm, and at its base the cell membrane becomes attached to the annulus. As the cytoplasm moves distally during development of the spermatid, the annulus maintains its position relative to the nucleus so that the tail lies in a flagellar canal, which is filled with a tongue of Sertoli cell cytoplasm.

Initially during spermiogenesis, the endoplasmic reticulum (ER) exists as a three-dimensional network of spherical and tubular cisternae connected by narrow tubules, especially near the Golgi apparatus and lining the plasma membrane. The network extends through the intercellular bridges joining the cell to other spermatids. There are particular modifications of the ER in the vicinity of the tail during its development. During the last stages of spermiogenesis the ER regresses and disappears.

Several strange evanescent structures appear and disappear during spermiogenesis. These include the manchette, a sleevelike structure extending into the cytoplasm from a nuclear ring which forms in the cytoplasm near the margin of the acrosome. There is also a chromatoid body which forms from nuage found among the mitochondria of the spermatocytes. For a time, it seems to interact with the nucleus before migrating to the end of the mid-piece and disappearing.

The mitochondria of the developing spermatid gradually assemble themselves around the mid-piece of the spermatozoon, once the manchette has disappeared, to form a very characteristic double helix between the neck and the annulus.

*Release of the spermatozoon.* The first step in the release of the spermatozoon is the breakdown of the tubulo-bulbar complex (see p. 80), which presumably acts as an anchoring device during the last stages of sperm development. Separation of the complex also serves to eliminate a large fraction, up to 70 per cent, of the spermatid cytoplasm, by transferring it to the Sertoli cell. The rest of the spermatid cytoplasm remains attached to the neck of the spermatozoon, and as the spermatozoon moves into the lumen, the mass of cytoplasm does not shift, so that the thin strand of cytoplasm joining the mass to the spermatozoon gets longer and longer. Eventually this breaks, leaving a small remnant, known as the cytoplasmic or kinoplasmic droplet, just behind the sperm head, and the spermatozoon is released into the lumen, as an independent cell. The cytoplasm left in the Sertoli cell crypt, known as the residual body, is still linked by cytoplasmic bridges to residual bodies from other spermatozoa of the clone, but the whole mass is phagocytosed by the Sertoli cell.

## Spermatogenic cycles and waves
### The cycle of spermatogenesis
One of the most remarkable facts about the seminiferous epithelium is that certain cells are only ever found in association with certain other cells. For example an $A_1$ spermatogonium, about to undergo its first synchronized division, is always found with preleptotene spermatocytes, pachytene spermatocytes, young round spermatids, and mature spermatids about to be released as spermatozoa. Each of these cells develops in synchrony with the others, so that if we could watch one section of the tubule wall with a time-lapse camera, a series of different cell associations would be seen, until the cycle was completed and the original cell association appeared again, but with each cell now one generation further on. The cycle must be repeated four times from the first synchronized division of an $A_1$ spermatogonium to the liberation of the spermatozoa derived from that particular cell, and the population of $A_1$ spermatogonia appears during the latter half of the preceding cycle (their origin is discussed on p. 83). The amount of tubule showing a particular cell association varies among species. In rats, considerable lengths, up to 10 mm, can be at the same stage, and complete cross-sections of tubules in sections of testes of most species

usually show the same cell association. However, in man, the area of tubular wall showing the same cell association is much smaller, so that most tubular cross-sections show several different associations. However, if allowance is made for the smaller areas involved, the same organized development can be discerned.

Table 4.1. *The relation between the various classifications used for dividing up the spermatogenic cycle.* The height of the section of each column corresponds approximately to its relative duration in the rat

| Simplified system | Parvinen's description of tubule under phase contrast | Leblond and Clermont's scheme | von Ebner, Curtis, Roosen-Runge and Ortavant's scheme |
|---|---|---|---|
| Elongation | Pale | IX | 1 |
|  |  | X |  |
|  |  | XI | 2 |
|  |  | XII |  |
| Grouping | Pale spots | XIII | 3 |
|  |  | XIV | 4 |
| Maturation |  | I | 5 |
|  | Dark spots | II |  |
|  |  | III |  |
|  |  | IV |  |
|  |  | V | 6 |
|  |  | VI |  |
| Release | Dark centre | VII | 7 |
|  |  | VIII | 8 |

Fig. 4.19. Cell associations involving always four or five generations of germ cells in the tubules of a rat testis during the development of a spermatozoon from an A₁ spermatogonium. Because the development of the different generations is synchronized, the same cell associations recur. The pictures represent the changes *with time* at one point in a tubule; the horizontal axis represents time in days, and the width of each section is proportional to the duration of that stage. The stages depicted are elongation (E), grouping (G), maturation (M) and release (R), so-called from the condition of the spermatids (the older generation, when there are two as in stages M and R), but these stages always have the same types of earlier cells in association with the particular type of spermatid. Note that the same cell associations recur with each complete cycle of 12 days, and with each complete cycle, the cell we are concentrating on (indicated by heavy outline and arrows) is one generation further ahead. There are four cycles from the division of the A₁ spermatogonia, three cycles from the beginning of the meiotic prophase and

one and two-thirds cycles from the meiotic divisions to the liberation of the spermatozoa. (For the origin of A₁ spermatogonia see Fig. 4.16.) The spermatogonia (A₁, A₂, A₃, A₄, In and B) divide in the basal compartment between the boundary tissue and the special junctions (shown here as a double line) between pairs of Sertoli cells. Then at the beginning of the meiotic prophase, either in preleptotene (PL) or leptotene (L), a new set of junctions forms underneath the germ cells, but as the old junctions persist for a while, the germ cells during this time are isolated in an 'intermediate' compartment (see also Fig. 4.17). With the start of zygotene (Z), the old junctions open up so that cells at this stage and during the rest of meiosis (pachytene (P) and diplotene (D)), as well as the early spermatids, are to be found in an adluminal compartment, in the intercellular spaces between pairs of Sertoli cells. Then as the older generation of spermatids is shed (arrows into lumen) the young spermatids move to the crypts in the luminal surface of the Sertoli cells, where they complete their development.

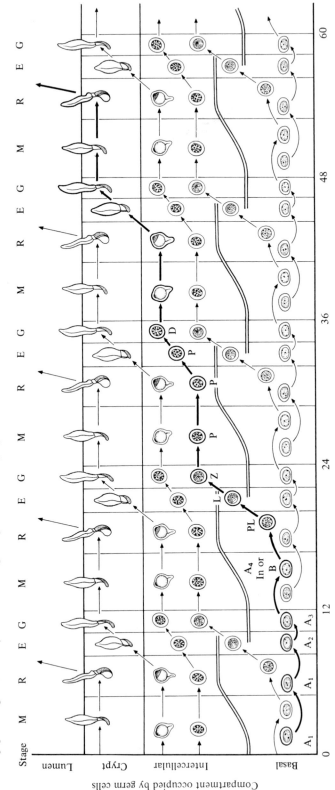

Several schemes are currently used to classify the various cell associations (Table 4.1). The Leblond and Clermont scheme based on the developmental state of the acrosome, recognizes 14 different stages in the rat, usually numbered I to XIV, beginning immediately after the meiotic divisions. The scheme used by von Ebner, Curtis, Roosen-Runge and Ortavant, based on the overall pattern of the cell associations, recognizes eight different stages in several species, usually numbered 1 to 8, beginning immediately after the release of the spermatozoa. Both schemes are probably too detailed for general use, and therefore I would like to suggest that there are four basic stages in each cycle: the post-release or elongation (E) stage, the pre-meiotic or group-formation (G) stage, the post-meiotic or maturation (M) stage and the pre-release (R) stage (Fig. 4.19). The other schemes can be easily reconciled with this simple EGMR classification, and – also important – we can recognize these four stages in isolated living seminiferous tubules of rats by examining them under a phase-contrast microscope. During the E stage, the tubules are pale and translucent; they develop a weak spotty pattern at the beginning of the G stage; the spots become darker and more obvious in the M stage, until a homogeneous dark centre to the tubule is seen in the R stage.

*The wave of spermatogenesis.* People are often confused between the spermatogenic cycle, which is in time, and the spermatogenic wave, which is in space. The cycle involves changes with time in the appearance of one section of a tubule. To see the wave you must travel along a tubule

Fig. 4.20. A diagram illustrating the wave of spermatogenesis within successive lengths of tubule, showing earlier stages of the cycle as one passes along the tubule from the rete. Also shown is one modulation (shaded) where a length at the G stage instead of the R stage follows a length of the E stage; thereafter the normal order resumes.

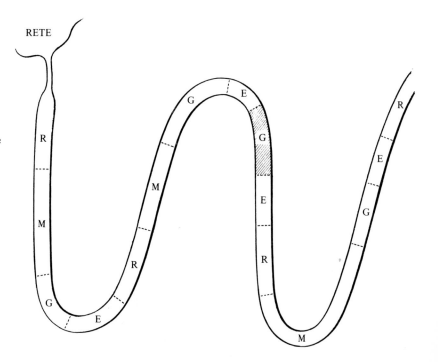

beginning at the rete; successive lengths of the tubule would be at earlier stages in the cycle, forming a wave made up of the spatial arrangement of stages in the temporal cycle. The section next to the rete may be at any stage but if, for example, it is at release stage (R), then the next section will be at the maturation stage (M), then a section at grouping stage (G), then one at elongation stage (E) and then one at the release stage (R) again, and so on (Fig. 4.20). The same sequence would be found regardless of where you started in the tubule, except that (*a*) at some points along the tubules there could be modulations or local reversals in the order, for example a series RMGE RMGE may be followed by G instead of R; the series then resumes with E RMGE R etc., and (*b*) as the tubule is two-ended, the two series must meet somewhere, even if there are no modulations, at a point known as the site of reversal. The wave is virtually impossible to discern in the human testis, as the areas present at any one stage of the cycle are so much smaller than in other species.

The cycle and the wave therefore have very little to do with one another. The cycle is a consequence of synchronized development of different types of germ cells. No one knows quite how the wave originates, but it must be a consequence of the way spermatogenesis begins at puberty, and, once established, it persists because of the synchronized development of different cells.

*Duration of spermatogenesis*

In a random series of tubular cross-sections, the longer-lasting stages of the spermatogenic cycle will occur more commonly if enough sections are examined. (This does *not* mean that the lengths of the different stages in a *single* wave will always be in proportion to the duration of those stages in time.) This principle has enabled estimates to be made of the relative durations of the various stages, as fractions of a whole cycle. The length of the whole cycle can be determined by using radioactive thymidine. This substance is incorporated into DNA during its synthesis, and the synthesis of almost all the DNA for the whole of meiosis occurs during pre-leptotene, during the release (or R) stage of the cycle. Therefore if the most advanced labelled cells are determined by autoradiography at various times after injection of thymidine, the rate of progression of cell development can be estimated. For example, if the most advanced labelled cells are pachytene spermatocytes in the R stage of the cycle, then one whole cycle has elapsed since the time of injection; if the most advanced labelled cells are young spermatids at the same stage, then two cycles have elapsed; if spermatozoa about to be released are the most advanced labelled cells, then three cycles have elapsed.

With these techniques, the length of the spermatogenic cycle has been found to be constant for any one species under a variety of conditions, ranging between 8 and 16 days for different species. The human is at the upper end of this range.

*Fertile life*

The first spermatozoa are shed from the testis at puberty, at a time when androgen production is still well below adult levels. The sperm production rate per testis continues to rise as the testis enlarges; the sperm production rate per unit weight of testis does not usually change at this time.

In old age, the incidence of degenerative conditions increases, but there is no conclusive evidence that testis size decreases with age in normal individuals, even in humans up to 80 years, mice up to 31 months, and rats and rabbits up to 3 years old. Individual humans are believed to have fathered children when well over 90 years old. However, there is some evidence that the numbers of spermatozoa in the testes decline in mice from about 12 months of age onwards, and there is a well documented decrease in the concentration of free testosterone in the blood of several species (see Book 3, Chapter 4).

## Biochemical aspects of spermatogenesis
### Synthesis of DNA and RNA
DNA is synthesized during each of the spermatogonial mitoses, but although the length of each cell cycle remains approximately the same, the S (synthesis) phase becomes progressively longer and the $G_2$ (second gap) phase correspondingly shorter. When the bulk of the DNA for the meiotic divisions is synthesized during preleptotene, the S phase is again prolonged. The synthesis of DNA by the X- and Y-chromosomes occurs after that by the autosomes.

During zygotene and pachytene there is a further low level of DNA synthesis, estimated in plants to be 0.3 and 0.1 per cent, respectively, of that during the pre-meiotic S phase. The DNA synthesized during zygotene is probably involved in the alignment of the homologous chromosomes, and that during pachytene for repair.

Both the X- and Y-chromosomes appear to be inactivated during the meiotic prophase, but ribosomal and non-ribosomal RNA are synthesized on the autosomes during this period. The synthesis of both forms of RNA rises to a peak in late pachytene and then falls again to very low values at the time of diakinesis. This pattern of RNA synthesis can be related to the formation of the lampbrush organization of the chromosomes, and also to the pattern of protein synthesis. The RNA formed during meiosis is unusually stable. It is chiefly heterogeneous and of high molecular weight; most of it is not transferred immediately to the cytoplasm but remains associated with the chromosomes.

RNA synthesis resumes again in young round spermatids but stops as the spermatids move from between the Sertoli cells to luminal crypts. The rate of RNA synthesis per unit DNA content is the same in round spermatids as in pachytene spermatocytes, and round spermatids also synthesize both ribosomal and polyadenylated RNA. However it is important to remember that a considerable fraction of the ribosomal and polyadenylated RNA produced by the pachytene spermatocytes is preserved until the later stages of spermatid development.

### Synthesis of proteins

Protein synthesis in spermatocytes follows RNA synthesis in rising to a peak at pachytene and then falling to negligible values at diakinesis. In spermatids, cytoplasmic protein synthesis occurs throughout spermiogenesis. Synthesis of nuclear protein occurs at a low level in early spermatids, then stops completely at about the time the spermatozoa move into the luminal crypts. Then, as the spermatozoa group together and for a little while after the next generation of spermatocytes divide, protein synthesis in the nucleus resumes again, but now involves the synthesis of arginine-rich basic proteins.

Testis-specific histones may be present in spermatocytes, together with histones also found in other tissues. During spermiogenesis in mammals, these histones are removed at the time of nuclear condensation, but are not replaced directly by protamines. Instead unusual basic proteins are transiently associated with the condensing spermatid nucleus, but as these proteins are not found in mature spermatozoa they should not be classed as protamines. The true protamines of mammalian spermatozoa are small basic proteins (some 50 amino acids, about half of which are arginine) which are synthesized during the final stages of spermiogenesis.

It is interesting that most of the soluble proteins of the spermatozoa, such as the proteins of the plasma membrane, the acrosome and mitochondrial matrices, are synthesized primarily during the meiotic prophase, whereas the 'structural' proteins of the tail are synthesized in the spermatids. A number of enzymes first appear in developing testes at the same time as the primary spermatocytes. These include sorbitol dehydrogenase, ribonucleotide phosphohydrolase, uridine diphosphate phosphohydrolase, $\alpha$-glycerophosphate dehydrogenase and carnitine acetyl transferase, and specific isoenzymes of acid phosphatase, hexokinase and lactate dehydrogenase. This last enzyme has been shown to be synthesized in the mouse only by pachytene spermatocytes and spermatids. A testis-specific cytochrome is also found in primary spermatocytes and spermatids.

### Haploid gene expression

The evidence available suggests that very little gene expression occurs in the haploid spermatids or spermatozoa. In other words, it seems extremely unlikely that the genotype of a spermatozoon would be revealed in either its structure or function. This is the obstacle that has so far defeated all attempts to separate X- from Y-bearing spermatozoa, and hence to influence the sex ratio at conception. Many claims for success have been made but none has been confirmed, and it is tempting to conclude that Nature has seen fit to protect the primary sex ratio from sperm selection, because of the confusion this could cause in the process of adaptive evolution.

Although we may be unable to separate living X-bearing from Y-bearing spermatozoa, it is tantalizing to discover that we can identify Y-bearing spermatozoa of humans after fixation and staining with quinacrine, when the Y-chromosome is revealed as a bright ultraviolet-fluorescent spot (see Fig. 4.21). It is only spermatozoa of humans that consistently fluoresce in this way, but at least it allows us to check sperm selection procedures in the laboratory, without having to resort to sexing large numbers of offspring before we can establish whether the procedure has worked.

There is, we should note, one clear exception to the 'rule' about haploid expression, and this has to do with the mutants at the *T* locus in the mouse. The dominant allele *T* causes short tails; in the homozygote it is lethal, but when combined with one or other of several recessive alleles $t^{1, 2, etc.}$, produces tailless animals. When heterozygous (*T/t* or +/*t*) males for some *t* alleles are mated, more than half the offspring produced carry the *t* gene, sometimes as many as 90 per cent, but this does not happen with heterozygous females. Similar numbers of the two types of spermatozoa are produced, but apparently the *t*-bearing spermatozoa have some advantage over the *T*-bearing or the +-bearing spermatozoa, and we feel compelled to infer that this is because of gene expression in the haploid cells. It appears that the *t*-bearing spermatozoa can survive better in the female tract, since the difference between the two types of spermatozoon disappears if mating is delayed for about 6 hours from its normal time, at about the time of ovulation. With mating at the normal time, there is an interval of 2 or 3 hours before fertilization begins: with delayed mating, the spermatozoa can start to fertilize the eggs almost immediately.

Moreover, the recent evidence that spermatids can synthesize both ribosomal and polyadenylated RNA shows that a mechanism does *exist* for haploid gene expression, although we do not know yet whether these

Fig. 4.21. Chimpanzee spermatozoa stained with quinacrine. The dye induces part of the Y chromosome to fluoresce, but unfortunately in this species, as in the gorilla, some of the autosomes also fluoresce, so the dye does not invariably reveal the presence of a Y chromosome. In human spermatozoa, however, fluorescence is Y-specific. (From H. Suánez, J. Robinson, D. E. Martin and R. V. Short, *Cytogenet. Cell Genet.* **17**, 317–26 (1976).)

products are actually translated to provide specific proteins. The carry-over of RNA from the spermatocyte through meiosis means that the diploid genome can also influence sperm development. One other complicating feature is the persistence of the cytoplasmic bridges between clones of spermatids, each of which can be presumed to have its own unique haploid genome. Any RNA passing from the nucleus into the cytoplasm of any individual spermatid should be able to pass through into the cytoplasm of other members of its clone, but no one knows whether this really happens or how important it is.

*Immunology of spermatogenesis*

We have long known that an animal can be immunized against its own spermatozoa. This does not normally happen because the antigens responsible are segregated behind the blood–testis barrier. Furthermore, if a systemic immune reaction is produced by injection of spermatozoa, there is not automatically a reaction of the antibody with the germ cells carrying the antigen because the barrier normally prevents the antibodies (and other molecules of that size) from reaching the germ cells in the adluminal compartments. However, adjuvant and spermatozoa injected at the same time produce an autoimmune orchitis and aspermatogenesis; this also often happens in the second testis of a normal individual if the first testis is damaged (see Book 4, Chapter 6).

Antigens similar to those on the spermatozoa can be found not only on all haploid cells in the testis but also on spermatocytes from pachytene onwards, i.e. from just after the germ cells pass into the adluminal compartment. The surface area of the pachytene spermatocytes is about four times that of cells in preleptotene, so the antigens are presumably some of the surface membrane synthesized during that period. The injection of pachytene spermatocytes with adjuvant will also produce immunological aspermatogenesis.

**Hormonal control of spermatogenesis**

Philip Smith showed more than 50 years ago that hypophysectomy (removal of the anterior pituitary gland) causes atrophy of the testes, a fall in the secretion of androgen to very low levels, and cessation of spermatogenesis. Shortly afterwards, it became known that the pituitary contains two gonadotrophins (usually called follicle-stimulating hormone (FSH) and luteinizing hormone (LH) from their actions in the female). In the male, LH stimulates the Leydig cells in the interstitial tissue to increase their androgen production (see Book 3, Chapter 4), and is therefore sometimes called interstitial cell stimulating hormone (ICSH). On the other hand, FSH has little or no direct effect on androgen production; it has been said to 'stimulate' the tubules. Complete spermatogenesis can only be restored by treatment with both FSH and LH or with FSH and testosterone in a rat whose testes have regressed after hypophysectomy. However, the situation is not straightforward because spermatogenesis can be maintained immediately after hypophysectomy with just large doses of testosterone. (The large doses are presumably necessary to produce concentrations of the hormone inside the testis equivalent to those resulting from the local production of much smaller amounts, but curiously several other non-androgenic steroids are also effective.) The same does not apply to the sheep, where FSH is necessary for the maintenance as well as for the initiation or re-initiation of spermatogenesis, and the evidence available, though not entirely clear-cut, suggests that monkeys are more like sheep than rats in this respect.

It is therefore perhaps unfortunate that almost all the studies on mechanism of action of FSH have been undertaken in the rat, and mainly in the immature (18- to 25-day-old) or hypophysectomized adult rat. In these animals, FSH certainly has a wide range of biochemical actions (see Book 7, Chapter 2) especially on the Sertoli cells. However, it cannot be stressed too strongly that these findings apply only to immature or hypophysectomized adult rats, and there is just no evidence for other species or indeed for the normal adult rat. That the action of FSH is restricted to the Sertoli cells (and possibly the spermatogonia) is not surprising as neither gonadotrophin crosses the blood–testis barrier to any appreciable extent and therefore would never reach the spermatocytes and spermatids. In contrast, testosterone not only crosses the barrier readily but indeed appears to be transported across it by a facilitated diffusion carrier. Nevertheless, there is good evidence that the major site of androgen action in the tubules is also on the Sertoli cells, which have definite androgen receptors as well as FSH receptors. The myoid cells may also respond to androgens as they also have androgen receptors, but evidence for androgen receptors on germ cells is contradictory. A direct action of androgens on germ cells seems unlikely as the gene for the receptor is on the X-chromosome, and this is inactivated during meiosis, or even, as in the field vole, actually lost from the developing germ cells. There is also the evidence of fertile chimaeric mice with androgen-responsive Sertoli cells and androgen-insensitive germ cells (see Book 7, Chapter 4).

There is no hormonal treatment known to increase sperm production in adult animals in which spermatogenesis is already fully established, although variation in the extensive degeneration that occurs at several steps in spermatogenesis may indicate possible control points. Furthermore, after unilateral castration compensatory hypertrophy of the spermatogenic function of the remaining testis does not seem to occur unless this is done in immature animals or out-of-season breeders.

Fig. 4.22. Histological sections of rat testes. (a) Normal tubule at the end of the maturation stage of spermatogenesis showing mature spermatids (Z) ready to be shed into the lumen of the tubule, young spermatids (T) peripheral to them, and then a layer of pachytene spermatocytes (P). Nearest the outside, there are some Sertoli cell (S) and spermatogonia (G) to be seen. (b) Tubule also at the maturation stage in a testis that had been irradiated 21 days before with 400 rad of γ-rays from a caesium source. This testis was reduced to about 1.3 g from the normal weight of 1.7 g, but both generations of spermatids (Z and T) are still present in approximately normal numbers. However, there are no spermatocytes because 21 days earlier these cells would have been dividing spermatogonia, which are particularly sensitive to irradiation. At later times (30–50 days), the spermatids would be missing for the same reason, but the tubules then would be beginning to show repopulation from the stem ($A_S$ or $A_O$) spermatogonia, which are resistant to X-irradiation. (c) A section of a testis that had been heated 21 days earlier by immersion for just 30 min in a water bath at 43 °C; the rest of the rat was at room temperature. Some tubules contain virtually no germ cells except spermatogonia; in others there are a few spermatocytes (C) but there are no spermatids, because this amount of heat completely destroys the pachytene spermatocytes which would have developed into spermatids in 21 days. Eventually, by about 60 days after the heating, the tubules will be repopulated because some of the spermatogonia are heat-resistant. (d) A section of a testis of a rat that had been fed a diet deficient in essential fatty acids for 46 weeks from 4 weeks of age. Most of the tubules in this section contain only Sertoli cells and germ cells; other tubules (e.g. lower right) have a few spermatocytes (C) and early spermatids (T), but there are no late spermatids. (Sections (a) and (b) were prepared after Bouin's fixative had been perfused at arterial pressure through the testicular artery. Sections (c) and (d) were fixed by immersion. The tissues were embedded in paraffin and the sections were stained with haematoxylin and eosin. Section (c) was supplied by Mr K. A. A. Galil and Section (d) by Dr W. M. F. Leat.)

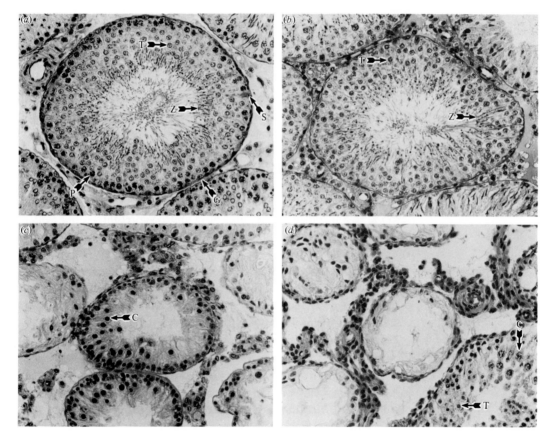

100 μm

## Abnormalities of spermatogenesis

### X-irradiation

Dividing spermatogonia are highly susceptible to radiation damage. As little as 300 rad given to the testes of rats will virtually eliminate them, but the stem cells, spermatids and spermatozoa are not obviously damaged. However the kinetics of the spermatogenic cycle do not change, so a 'maturation depletion' moves through the epithelium at the same rate as the missing cells would have done, and so successive cell types are missing from the epithelium (Fig. 4.22). The spermatogonial population is re-established from the stem-cell or $A_0$ spermatogonia, and the epithelium is repopulated as these cells progress through spermatogenesis. Higher doses of radiation will damage or kill more advanced cell types as well.

The male offspring of pregnant rats that have been given even lower doses of whole-body radiation (about 130 rad) on day 19 of pregnancy completely lack germ cells, which is a little strange as the germ cells are not dividing actively at that time.

### Heat and cryptorchidism

If, in mammals that normally have scrotal testes, a testis does not descend into the scrotum, or is returned experimentally to the abdomen, spermatogenesis does not proceed. This appears to be due to the higher temperature in the abdomen, because if a cryptorchid testis is cooled, but left in the abdomen, spermatogenesis will proceed normally. After experimental cryptorchidism, histological changes appear within 2 days and only Sertoli cells and spermatogonia remain after 2 weeks.

Similar effects can be produced by insulating the scrotum or by immersing the testes in warm water. In rats, exposure of the testes for 30 min to a temperature of 43 °C (rather cooler than most people's bath water) results in a halving of testis weight and profound disturbances in spermatogenesis (see Fig. 4.22). Lower temperatures are also effective if applied for longer times, and higher temperatures require even shorter exposures. The most sensitive cells seem to be pachytene spermatocytes, with the early spermatids next; again 'maturation depletion' follows a single treatment. The spermatozoa in the epididymis seem to be relatively resistant. One cannot help wondering whether fertility in men could be reduced by the wearing of very tight trousers, and concern about the possible ill-effects of sauna baths has stimulated a considerable amount of research in Finland; no deleterious consequences have yet been convincingly demonstrated for either circumstance.

No one knows why the testis cannot function at body temperature in animals that normally have scrotal testes. Hypoxia induced by an increase in metabolism without a corresponding change in blood flow has been suggested, but this is certainly not the whole explanation as the testis can withstand complete interruption of its blood flow for at least 30 min without any obvious signs of damage.

*Dietary deficiencies*

The spermatogenic function of the testis is depressed if food intake is severely reduced, although the production of androgens is affected to a greater extent. These changes can largely be explained by changes in gonadotrophin secretion by the pituitary. The testis also degenerates in a number of specific dietary deficiencies, including those of vitamin A, vitamin E (in rats), essential fatty acids (see Fig. 4.22), certain amino acids and zinc, but in these conditions, deficiency appears to influence the testis directly; there is evidence that pituitary function is normal, or even increased, as a result of decreased negative feedback from the testis. Vitamin B deficiency seems to affect the pituitary and, indirectly, the testis.

*Chemical damage*

The testis is damaged by a wide range of chemicals. Probably the most notable effects are seen with cadmium salts; doses that affect no other tissues in the body, cause severe necrosis in the testis. As one might expect, cytotoxic drugs such as Busulphan (butane dimethanesulphonate), methotrexate, vinblastine and vincristine used for treatment of malignancy are also toxic to spermatogonia, and the rest of the epithelium is then lost by maturation depletion. Hydroxyurea, which interferes with DNA synthesis, has a similar effect. Other drugs originally introduced for other purposes, such as stopping the growth of bacteria or killing schistosomes or amoebae, sometimes also cause specific damage to the testes of those taking the drug. The nitrofurans, niridazole, and bis-dichloracetyl-diamines are examples of these three types of drugs. Ethylene dibromide, used for fumigating grain, and dibromochloropropane used to kill soil nematodes, have both been incriminated as causing infertility by an action on the testes of farm animals and men.

However the most common abnormality of spermatogenesis in humans is a general reduction in spermatogenesis associated with a reduced number of spermatozoa in the semen (oligozoospermia). The cause of this condition is unknown, and as it is by no means unusual, its prevalence should encourage us to continue investigating the complex processes involved in the production of spermatozoa.

*Other causes*

The testis is deranged in a number of genetic abnormalities (see Book 2, Chapter 3). Also many men are infertile following mumps orchitis, but we do not know whether this is a direct effect of the infection, a consequence of the inflammation and increased temperature of the testis, or an immunological reaction (see Book 4, Chapter 6), following breakdown of the blood–testis barrier.

Spermatogenesis is a most intricate and involved process, but this is not surprising when one considers the remarkable degree of specialization of the spermatozoon. Its genetic material is only half that of somatic cells, but it has acquired many attributes that they do not possess. This nuclear weapon must travel great distances to reach its target, and in doing so it must penetrate several defence systems for which it is uniquely adapted both morphologically and biochemically.

### Suggested further reading

The molecular biology of mammalian spermatogenesis. A. R. Bellvé. *Oxford Reviews of Reproductive Biology* **1**, 159–261 (1979).

Summary and synthesis of a workshop on chromosomal aspects of male sterility in mammals. P. de Boer and A. G. Searle. *Journal of Reproduction and Fertility* **60**, 257–65 (1980).

Sites of action of androgens and follicle stimulating hormone on cells of the seminiferous tubule. I. B. Fritz. In *Biochemical Actions of Hormones*, vol. 5, pp. 249–81. Ed. G. Litwack. Academic Press; New York and London. (1978).

Biochemistry of male germ cell differentiation in mammals: RNA synthesis in meiotic and postmeiotic cells. V. Monesi, R. Geremia, A. D'Agostino and C. Boitani. *Current Topics in Developmental Biology* **12**, 11–36 (1978).

Sertoli–germ cell interrelations: a review. L. R. Russell. *Genetic Research* **3**, 179–202 (1980).

The functional significance of the blood–testis barrier. B. P. Setchell. *Journal of Andrology* **1**, 3–10 (1980).

*Effects of Hormones, Drugs and Chemicals on Testicular Function.* A. G. Davies. Eden Press Annual Research Reviews, vol. 1 (1980).

*The Spermatozoon.* Ed. D. W. Fawcett and J. M. Bedford. Urban and Schwarzenberg; Baltimore (1979).

*Frontiers in Reproduction and Fertility Control.* Ed. R. O. Greep and M. A. Koblinsky. MIT Press; Cambridge Mass. (1977).

*Handbook of Physiology, Section 7 Endodrinology, vol. 5, Male Reproductive System.* Ed. D. W. Hamilton and R. O. Greep. American Physiological Society; Washington, D.C. (1975).

*The Testis, vol. 1: Development, Anatomy and Physiology; vol. 2: Biochemistry; vol. 3: Influencing Factors.* Ed. A. D. Johnson, W. R. Gomes and N. L. Vandemark. Academic Press; New York and London (1970).

*The Testis, vol. 4: Advances in Physiology, Biochemistry and Function.* Ed. A. D. Johnson and W. R. Gomes. Academic Press; New York and London (1977).

*The Process of Spermatogenesis in Animals.* E. C. Roosen-Runge. Cambridge University Press (1977).

*The Mammalian Testis.* B. P. Setchell. Elek; London (1978).

*Testicular Development, Structure and Function.* Ed. H. Steinberger and E. Steinberger. Raven Press; New York (1980).

CARL A. RUDISILL LIBRARY
LENOIR RHYNE COLLEGE

# 5

# Sperm and egg transport

*MICHAEL J. K. HARPER*

In this chapter we will be concerned with the varied experiences of spermatozoa after they leave the seminiferous tubules and during their passage through the male and female genital tracts. These include changes that spermatozoa undergo up to the moment of fertilization, but the processes intimately connected with the acquisition of fertilizing ability, such as modifications in the sperm cell membrane during passage through the epididymis, and the events of capacitation and the acrosome reaction, will be reserved for treatment in Chapter 6 in which fertilization itself is dealt with.

As we shall see, spermatozoa and eggs usually meet and fertilization ensues in about the middle third of the Fallopian tube, a region referred to as the site of fertilization. Nearly always the spermatozoa reach this point hours (sometimes many hours) before the eggs get there, and so it seems appropriate to consider first the problems associated with sperm transport.

## Sperm transport in the male tract

### Mechanisms

After they leave the seminiferous epithelium in the testis, the spermatozoa, which at this time are almost completely immotile, are moved passively from the testis by the current of fluid secreted by the Sertoli cells, and pass into a branched reservoir, the rete testis, into which both ends of each testis tubule open. Efferent ducts (vasa efferentia) link the rete testis with the epididymis (Fig. 5.1). There are usually some 10 to 20 such ducts, which arise from extratesticular tissue near the upper pole of the testis. They are initially only slightly convoluted but become much more so as they reach the epididymis. Sperm transport from the rete testis into these efferent ducts seems to be aided by the reabsorption of water in the ducts, which tends to pull spermatozoa in that direction. Once in the ducts, spermatozoa are shifted by cilia lining the epithelium, which beat towards the epididymis, and by the contractile activity of the smooth muscle in the wall of the ducts.

The epididymis consists of a single, long, tightly convoluted duct with supporting connective tissue elements; it covers the posterior border of the testis, with its tail attached to the base of the scrotum by the genito-inguinal ligament. Conventionally, the epididymis is divided into three regions, the caput (head), the corpus (body) and the cauda (tail), but the exact boundaries defining them are indistinct and the function of the various sections can vary between species. For instance, in man much of the caput region is really part of the extensive efferent duct system. In nearly all species the cauda epididymidis is a bulbous terminal structure, in which the tubule becomes enlarged and packed with spermatozoa. Again this is different in man, whose cauda is inconspicuous.

Sperm transport inside the epididymis is the result of spontaneous, peristalsis-like contractions of the smooth muscle lining the wall of the organ. Hydrostatic pressure inside the tubule may also play some role in sperm movement. In some species, such as the guinea pig, pressure is high in the caput region and falls progressively to the first part of the cauda. Hydrostatic pressure is again increased in the last part of the cauda, which is a bit of a mystery for there is no good anatomical reason why this increased pressure is not more evenly distributed back through the corpus. At least from the caput to the first part of the cauda the spermatozoa are moving down a pressure gradient.

Spermatozoa are transported from the epididymis into the vas deferens chiefly by the steady flow of secretion through the epididymis. The vas

Fig. 5.1. Human testis and epididymis, showing efferent ducts leading from the rete testis to the head or caput of the epididymis, and the tail or cauda of the epididymis continuing to become the vas deferens. (From M. Dym. *Histology*, 4th edn, Ed. L. Weiss and R. O. Greep, p. 981. McGraw-Hill; New York (1977).)

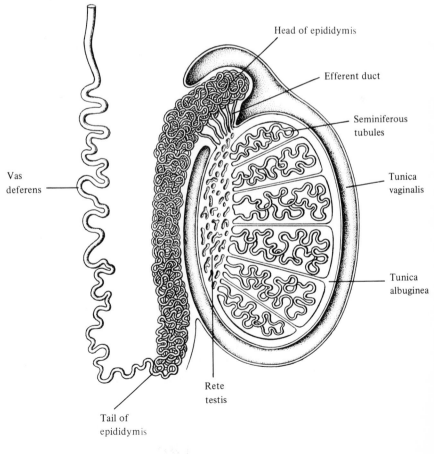

Head of epididymis

Efferent duct

Seminiferous tubules

Vas deferens

Tunica vaginalis

Tunica albuginea

Rete testis

Tail of epididymis

deferens is a long tube with a thick muscular wall which in some species has a distended portion called the ampulla (Fig. 5.2). During the process of ejaculation powerful peristaltic contractions propel the spermatozoa along the two vasa deferentia to the ampullae and thence into the urethra, which is a single duct extending the length of the penis. Their removal then permits more spermatozoa to move down from the corpus, the content of which is replenished from the efferent ducts and caput. Thus frequency of ejaculation can affect the rate of sperm transport through the cauda. The time required for sperm transport from the caput to the cauda is fairly constant at 3–5 days in a variety of species. The total sperm-transport time from the seminiferous tubule to the exterior in sexually active men is about 10–14 days.

*Composition of epididymal fluid*
Sperm concentrations in the luminal fluids of the seminiferous tubules and the rete testis are low, but increase as the spermatozoa traverse the epididymis in most species. This is due to reabsorption of water by the epithelium of the efferent ducts and caput epididymidis. In some animals such as the bull, hamster and probably man, reabsorption of water takes place mainly (over 90%) in the efferent ducts, so that sperm concentrations do not increase significantly along the length of the epididymis. In the rat, where precise measurements have been made, sperm concentrations in the caput epididymis are $0.66 \times 10^9$ sperm/ml, and increase steadily to

Fig. 5.2. Lower part of the trunk in a man, showing the reproductive tract and neighbouring organs. A, ampulla; B, bladder; Bu, bulbo-urethral or Cowper's glands; E, epididymis; P, prostate; Sv, seminal vesicle; T, testis; U, urethra; Vd, vas deferens.

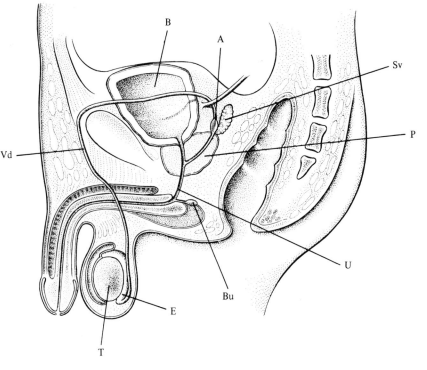

$1.84 \times 10^9$ sperm/ml in the cauda. In association with this trend there are changes in the concentrations of some inorganic ions. Potassium rises steadily from caput to cauda (20.5 to 40.0 mM), while sodium falls (104 to 37 mM). In other species the increase in potassium ions along the length of the epididymis is not so distinct, or even consistent. Diminishing concentrations from caput to cauda are seen for calcium (0.85 to 0.25 mM), but concentrations of magnesium (3 mM) and sulphur (31 mM) are highest in the corpus. Chloride concentration, by contrast, is stable throughout the epididymis (about 27 mM). The phosphorus content is high in the epididymis compared to serum (94 versus 2 mM, respectively) but shows no distinct trend.

Rete testis fluid is iso-osmolar with blood serum, but seminiferous tubule fluid and all epididymal fluids are hyper-osmolar. In the rat, at least, the osmolarity of epididymal fluids declines from the caput to the cauda. The reason for this is not clear, since sodium is being steadily removed; other molecules must move in to maintain the hyper-osmolarity. Only 32 per cent of the total osmolarity of the caput fluid and 18 per cent of the cauda fluid can be accounted for by electrolytes. Of the remaining osmolarity, 30 per cent is due to two organic compounds found almost exclusively in epididymal fluid: glycerylphosphorylcholine and carnitine. The substances contributing to the remaining osmolarity are unknown, and the mechanism by which the hyper-osmolar state is maintained remains an enigma, especially since the epithelium is permeable to water and fluid flow is slow.

## Physiological changes in spermatozoa

We have long suspected that spermatozoa must undergo important though subtle changes while they traverse the male reproductive tract, because they acquire the art of motility and the potential to take part in fertilization. Spermatozoa recovered from the rete testis show only a weak vibrating movement of the tail, and this is followed in the caput of the epididymis by a slow circular swimming movement. By the time the spermatozoa have reached the cauda this has changed to a vigorous unidirectional progression with longitudinal rotation of the head. Such changes have been observed in several species, e.g. rat, rabbit and guinea pig; in the human, the pattern of motility does change during epididymal transit but it is not so clearly defined – the circular type of activity does not occur and spermatozoa recovered from the caput of the epididymis can exhibit immotility or slight vibratory activity of the tail, or even a wild thrashing movement. Rapid forward progression is seen in only a few spermatozoa in the middle of the corpus and in most spermatozoa in the cauda of the epididymis and in the vas deferens. Acquisition of progressive motility seems to be more a function of the ageing of the spermatozoa than of the microenvironment of particular regions of the epididymis. In the rabbit, it has been found that spermatozoa can become strongly motile in the corpus of the epididymis if their onward passage is prevented by ligation, but such spermatozoa are not capable of fertilizing eggs. Though we have no definite information for other animals, there is the important implication that motility is not necessarily an indication of fertilizing ability.

Although rabbit spermatozoa seem to acquire fertilizing capacity only after transit through the caput and a major portion of the corpus of the epididymis, there is some indication that human spermatozoa may be different. In men whose vas deferens has been surgically joined on to the caput region of the epididymis (epididymovasostomy) because of a pathological blockage further down, fertility is sometimes apparently restored, which suggests either that epididymal maturation of the sperm-atozoon is not important in man or that it can occur in the vas deferens.

### Structural changes in the spermatozoa

During transit through the epididymis there are no obvious morphological changes in the tail of normal spermatozoa, but a cytoplasmic droplet severed from the residual body of the spermatid at the time of spermiation is first observed around the neck of the spermatozoon and slowly migrates caudally down the mid-piece; this droplet contains membrane-bound vesicles and other elements. The cytoplasmic droplet is usually lost by the time of ejaculation; its function is obscure, but experiments have shown that migration of the droplet is not essential for acquisition of motility.

Changes take place in the sperm head during epididymal transit and these include, in many species, alterations in the shape and size of the acrosome. These have not been observed in human spermatozoa, but they have in monkeys; their significance is not known. The sperm plasma membrane has been found to undergo maturational changes, as detected by observations on light-reflecting properties, uptake of vital dyes, binding of lectins, adhesiveness, net charge and charge density, and on auto-agglutination patterns (head-to-head or tail-to-tail). These things all seem to bear upon the sperm's growing potential for fertilization, and are discussed in Chapter 6.

### Role of hormones in sperm maturation

The ability of the epididymis to maintain sperm viability and permit maturation appears to be dependent on androgenic hormones. Hypophy-sectomy or treatment with anti-androgens can cause a rapid loss of fertilizing capacity of spermatozoa in the cauda of the epididymis of the rabbit, without necessarily inducing any morphological changes in the tissues. The androgenic hormones, testosterone and $5\alpha$-dihydrotestos-terone, presumably exert their action mainly by maintaining the secretory activity of the epididymal epithelial cells, but a direct action of androgens on the spermatozoa cannot be excluded. Androgen is carried to the rete testis by an androgen-binding protein (ABP) with an apparent mol. wt of 86 000–91 000. This ABP is of Sertoli-cell origin. As a consequence, rete testis fluid always contains androgens in greater concentrations than arterial blood, though in lesser concentrations than testicular venous blood, and therefore provides a good source of androgen for the caput epididymidis. There is some disagreement as to whether it is the blood androgens or those in rete testis fluid that are responsible for maintaining sperm viability and permitting maturation, but we have no doubt that androgen from whatever source is vital if spermatozoa are to acquire fertilizing ability in the epididymis.

Spermatozoa that are not expelled to the exterior by ejaculation are probably voided in the urine, and only a minute proportion are thought to be resorbed by the male genital tract.

**Seminal emission and ejaculation**
The adrenergic system is very important in seminal emission. Smooth muscle fibres of the male reproductive tract receive a large part of their sympathetic innervation in the form of 'short adrenergic neurons', i.e. post-synaptic neurons. Although parasympathetic fibres innervate the male tract, they probably do not participate in the motor innervation of the vas deferens, which is controlled by the α-adrenergic system. It is well known that if patients are treated for testicular tumours by extensive surgical removal of lymph glands in the thorax and abdomen, which un-avoidably involves removal also of sympathetic nerves from T12 to L3, seminal emission can no longer occur, not even retrogradely into the bladder, although erection is normal. Such patients experience a 'dry orgasm'. Treatment with agents that interfere with the α-adrenergic system can produce a similar effect in some individuals. Administration of the agent guanethidine, which is used in the treatment of high blood pressure, causes rapid loss of catecholamines from the short adrenergic neurons of the male reproductive tract, and if given chronically severely damages the neurons. As a consequence seminal emission fails but again potency and libido are unaffected. This evidence convincingly shows that emission is controlled by the activation of the smooth muscle of the vas deferens and urethra by noradrenaline release.

In contrast to seminal emission, the propulsive force in ejaculation probably comes from rhythmic contractions of two striated muscle groups, the ischiocavernosus and the bulbospongiosus, which surround the urethra in the pelvis and the penis (see Book 8, Chapter 2).

At ejaculation the spermatozoa, suspended in secretions from the testis and epididymis, are moved rapidly along the vas deferens into and through the urethra, being joined *en route* by the secretions of the accessory sex glands: the ampullary glands, the seminal vesicles, the prostate, and the bulbo-urethral (Cowper's) and urethral (Littré's) glands. All these secretions make up the fluid portion or plasma of the semen, but the relative contributions from different sources differ greatly and there are also large differences between animals. In man the bulk of the seminal plasma comes from the seminal vesicles (60 per cent) and prostate (30 per cent). The dog and cat, however, have large prostates but lack seminal vesicles, the pig has no ampullae and the bull has large seminal vesicles and only a small prostate. As Thaddeus Mann has made clear, these secretions contain distinctive chemical components so that the proportions coming from different glands can be assessed by analysis. Prostatic fluid contains citric acid, acid phosphatase and the metals zinc and magnesium, while seminal vesicle secretion is rich in fructose and in many species the long-chain fatty acids known as prostaglandins. The average pH of semen is basic, between 7.2 and 7.8, reflecting the relative contributions of acidic prostatic fluid and alkaline secretion from the seminal vesicles.

The usual volume of the ejaculate in a fertile man is about 2–6 ml, with a concentration of anything from 40 to 250 million spermatozoa/ml. However, 40 million should not be taken as the lower limit of fertility, as recent reports indicate that sperm counts as low as 10 million/ml (or 25 million per total ejaculate) may be normal in some fertile males. Furthermore, men taking large doses of androgens in an attempt to depress their sperm counts for contraceptive purposes were still fertile even though their sperm counts were reduced to as little as 1 million/ml. However, the presence of many immotile or abnormally shaped spermatozoa in the ejaculate is usually a good indication of male infertility.

At the time of ejaculation in man and other primates the seminal plasma

Fig. 5.3. A fresh ejaculate from a chimpanzee. It consists largely of coagulum, the small amount of fluid seminal plasma being visible in the narrow part of the tube. (From J. D. Roussel and C. R. Austin. *J. Inst. Anim. Tech.* **19**, 22 (1968).)

is either already coagulated or else coagulates within 1 minute after emission (Fig. 5.3). Proteolytic enzymes in the seminal plasma, which originate mainly from the prostate, then break down this coagulum causing liquefaction within 20–30 minutes. Full liquefaction should occur within 60 minutes in the normally fertile man. Complete motility of the spermatozoa is not achieved until liquefaction has occurred. Curiously enough, though, the first fraction of the ejaculate, which is sperm rich, can deposit spermatozoa that reach the cervical canal before coagulation occurs, and so spermatozoa do not all necessarily remain in the coagulum until liquefaction. Perhaps the function of the coagulum is to hold a reserve of spermatozoa which are released over a period of time. Coagulation occurs in rodents too, but the mechanism is different: a copulatory plug is formed in the vagina, after the spermatozoa have been deposited in the uterine lumen. But in bulls and rams, for example, the ejaculate never coagulates.

It is noteworthy that sperm concentrations in the ejaculates of the bull, ram, rhesus monkey, and others, can be of the order of $10^9$ sperm/ml which is much higher than in humans. The quantity of ejaculate also varies greatly from as little as 0.1 ml in the mouse to over 100 ml in the boar. Coitus is very prolonged in camels (24 hours!) and pigs, but very rapid in rats, rabbits and bulls.

## Sperm transport in the female tract
### Site of semen deposition
In some species (notably the rodents, pig and horse) spermatozoa ejaculated at coitus are deposited directly into the uterus, while in others, including the rabbit and primates, the spermatozoa are deposited into the vagina close to the external os of the cervix. By contrast with the seminal pH of 7.2–7.8, the vagina is very acid, being about pH 5.7 in women. The vaginal secretions therefore provide a rather hostile environment for the ejaculated spermatozoa, which do not remain viable there for long.

Even with vaginal deposition, some spermatozoa may be propelled directly into the cervical canal by the process of ejaculation, since they have been found in the Fallopian tube within 5 minutes in several species, including the human. The remainder of the spermatozoa trapped within the coagulum of seminal plasma must await liquefaction before contractions of the vaginal wall can cause a proportion to come into contact with the cervical mucus and thus give them a chance to ascend the female tract. Generally speaking, in those species in which spermatozoa are deposited in the vagina, the percentage that reaches the uterus is small. In other species, in which deposition of semen at ejaculation is into the uterine lumen, the cervical canal and its distinctive mucus do not form a barrier to the transport of spermatozoa to the site of fertilization at the ampullary-isthmic junction of the Fallopian tube. Table 5.1 shows numbers of spermatozoa ejaculated by different species and the site of deposition.

What is especially noteworthy about these figures is that despite the enormous differences in the total numbers of spermatozoa in the ejaculate, and the differences in the place of deposition, the actual numbers reaching the site of fertilization are remarkably similar and represent only a small fraction of those ejaculated. In those species in which ejaculation is into the vagina, little or no seminal plasma seems to enter the uterus.

*The cervix uteri*

A canal with very thick walls consisting mostly of connective tissue connects the vagina to the uterus; this is the cervix uteri. The canal is a complex affair, following an irregular course and with numerous deep crypts in the walls (Fig. 5.4). The mucous membrane lining the cervical canal is richly provided with glands that produce a copious secretion. Cervical mucus is composed of two main elements: cervical mucin and soluble components. Mucin is a glycoprotein, rich in carbohydrates, with a fibrillar system of long molecules linked either directly by disulphide bridges or indirectly by polypeptides. The soluble components of the mucus include inorganic salts, such as NaCl, and organic compounds of low molecular weight, such as glucose, maltose, mannose, and amino acids, as well as peptides, proteins and lipids. The soluble proteins are dispersed in the aqueous phase of the cervical mucus gel, which changes its viscosity with changes in blood levels of ovarian hormones. Under oestrogen dominance the mucus becomes thin and watery and the macromolecular fibrils are oriented in parallel chains or micelles, which permit spermatozoa

Table 5.1. *Species variation in number of spermatozoa ejaculated and site of deposition in different animals*

| Animal | Average no. sperm/ ejaculate (millions) | Site of sperm deposition | No. sperm in ampulla of Fallopian tube |
|---|---|---|---|
| Mouse | 50 | Uterus | < 100 |
| Rat | 58 | Uterus | 500 |
| Rabbit | 280 | Vagina | 250–500 |
| Ferret | — | Uterus | 18–1600 |
| Guinea pig | 80 | Vagina and uterus | 25–50 |
| Cattle | 3000 | Vagina | Few |
| Sheep | 1000 | Vagina | 600–700 |
| Pig | 8000 | Uterus | 1000 |
| Man | 280 | Vagina | 200 |

Data from various sources.

to swim through the mucus. Conditions at this time also favour sperm motility and this motility is thought to play an important role in the passage of spermatozoa through the cervix. In the luteal phase of the cycle, when progesterone is dominant, the mucus becomes viscous and rubbery with no apparent micellar structure, and thus is unfavourable for sperm penetration.

Other factors involved in sperm transport through the cervix besides sperm motility include muscular activity in the vaginal and cervical walls, the structure of cervical mucus, and the presence of the cervical crypts into which spermatozoa can enter. There are thought to be three phases of sperm transport through cervical mucus, at least in the human. There is an initial rapid phase in which relatively few spermatozoa reach the Fallopian tube; these are the ones that arrive within 5 minutes of coitus. This rapid phase of transport is then followed by a second or colonization phase in which large numbers of spermatozoa enter the crypts of cervical glands and remain there for a prolonged period of time. These glands could act as reservoirs that permit sperm release during the third and final prolonged phase of transport.

Survival of spermatozoa in the cervical crypts is prolonged, due to the fact that cervical mucus provides a physiologically favourable environment and somehow protects the spermatozoa from phagocytic attack. Observations made *in vitro* show that spermatozoa can traverse oestrous mucus at a rate of 0.1–3.0 mm/min. (Probably only in passage through cervical mucus, in negotiating the uterotubal junction and during actual penetration of the egg coverings does sperm motility play a significant role in

Fig. 5.4. Longitudinal section of the cervix in a cow. The deep and much branched crypts in the mucous membrane are clearly seen. E, external and I, internal openings (ostia) of the cervical canal; M, mucous membrane; S, submucous tissue. (From E. S. E. Hafez. In *The Biology of the Cervix*, ed. R. J. Blandau and K. S. Moghisshi. University of Chicago Press (1973).)

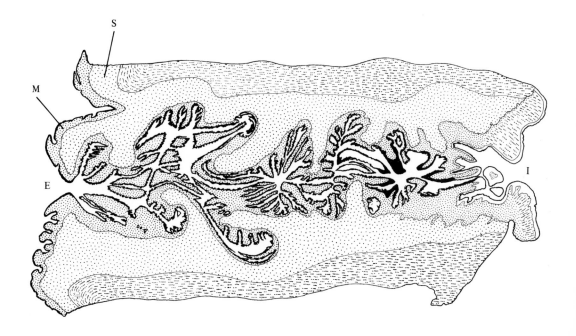

transport.) The closely packed long glycoprotein molecules of the mucus may well orient the spermatozoa during this part of their journey.

Sampling cervical mucus at different times after ejaculation shows that sperm concentrations in the upper portion of the cervical canal increase progressively with time. Nonetheless, only a small proportion of the spermatozoa actually reaches the uterus. In experiments in rabbits in which about 280 million spermatozoa were deposited in the vagina at ejaculation, only 600 000 (0.2 per cent) were found in the uterus 2 hours after mating and this percentage increased to a maximum of 0.6 per cent at 12 hours after mating (which is about the time of fertilization). The maximum number of spermatozoa found in the uterus of any one animal in these experiments was only about 1.5 per cent of those ejaculated.

### Spermatozoa in the uterus

Sperm transport through the uterus is probably fairly rapid, since as shown in Table 5.2 they can reach the Fallopian tubes within a few minutes in most species. Diane Settlage and her colleagues showed that in human subjects between 4 and 53 spermatozoa per tube could be found within 5–45 minutes after deposition of semen onto the external opening of the cervix by artificial insemination. This rapid transit must be mainly due to the vigorous contractions of the uterus, since spermatozoa are quite unable to swim the required distance in the available time, especially as they cannot be expected to move steadily in the direction of the Fallopian tube for there is no known mechanism to guide them. In addition, experiments have shown that non-motile objects (carbon particles, dead spermatozoa) and fluids (radiopaque solutions) are shifted quickly from one end of the uterus to the other. Uterine motility may be stimulated by the presence in seminal plasma of agents capable of causing smooth muscle contractions,

Table 5.2. *Time between coitus or artificial insemination and arrival of spermatozoa in the Fallopian tube*

| Animal | Time | Region of tube |
| --- | --- | --- |
| Mouse | 15 min | Ampulla |
| Rat | 15–30 min | Ampulla |
| Hamster | 2–60 min | Ampulla |
| Rabbit | few min | Ampulla |
| Guinea pig | 15 min | Ampulla |
| Dog | 2 min–few h | Oviducts |
| Sow | 15 min | Ampulla |
| Cow | 2–13 min | Ampulla |
| Ewe | 6 min – 5 h | Ampulla |
| Woman | 5–68 min | Oviducts |

Data from various sources.

notably the prostaglandins, or by the release of oxytocin from the posterior pituitary gland, which probably occurs in the females of some species, notably in women and cows, at the time of coitus.

Sperm survival in the uterine lumen is short, since phagocytosis by leucocytes begins a few hours after mating, and in any case much of the secretion and suspended matter in the uterus is expelled through the cervix. The uterine environment can also be hostile – of large numbers of spermatozoa placed in rabbit uteri (with both ends ligated) only 20 per cent could be recovered 6 hours later. On the other hand, live spermatozoa have been found in the human uterus 24 hours after coitus; very few were detected later, though many were still to be seen in the cervical mucus. The same survival period was noted in the rabbit, so that uterine survival is evidently similar in these dissimilar species.

Spermatozoa that manage to sequester themselves in uterine gland lumina (Fig. 5.5) possibly succeed in finding protection from phagocytosis and other lethal agents. In a number of different animals spermatozoa have been found in this location long after coitus, but it was not established whether they were still fertile or even alive. In certain bats, uterine spermatozoa are well known to maintain full fertility for several months, and recent observations suggest that this may possibly happen as a rare event in other species. Suzanne Ullmann, for instance, has produced good evidence that mice of a certain strain can experience two successive pregnancies from one series of copulations. Precisely in what part of the genital tract the mouse spermatozoa managed to survive is still a mystery; bat spermatozoa apparently see it through in the uterus and in the isthmus of the Fallopian tube (Fig. 5.6).

Fig. 5.5. Section of the uterus of a greater horseshoe bat *Rhinolophus ferrumequinum* showing spermatozoa packed into a uterine gland. The animal was collected during hibernation. (From C. R. Austin. *J. Reprod. Fert.* **1**, 151 (1960).)

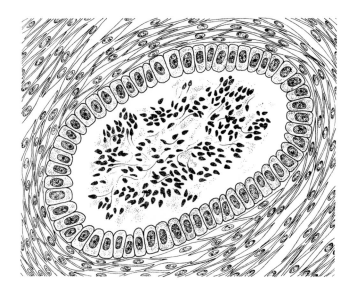

### Uterotubal junction

Interestingly, the number of spermatozoa ejaculated does not appear to influence greatly the number entering the Fallopian tubes, at least until very low figures are reached. In rabbits, mating once with a fertile buck (with an estimated 200 million spermatozoa deposited in the vagina), or artificial insemination of 21 million spermatozoa into the vagina, gave rise to very similar numbers (5000–6000) in the tubes. But insemination with only 0.5 per cent of the normal ejaculate (1.21 million) led to a reduction of sperm numbers in the tubes 1 day later, an average of only 88 being found.

The role played by the uterotubal junction in limiting passage of spermatozoa into the Fallopian tube is not clear, although in species such as the rat and hamster it forms a very efficient barrier. In these species millions of spermatozoa are ejaculated directly into the uterine horns and yet only a few thousand enter the tube. The anatomical complexity of the uterotubal junction varies with the species, being quite simple in some and highly complex in others (Fig. 5.7). It is difficult to force liquid or gas through this junction except at oestrus, or at the time when eggs are passing into the uterus, and thus it may act as a valve under hormonal control. There is still disagreement as to whether dead (non-motile) spermatozoa are usually transported into the Fallopian tube, although Halina Krzanowska has clearly shown that, in the mouse, morphologically abnormal spermatozoa are selectively precluded from entry into the Fallopian tube from the uterus.

### Spermatozoa in the Fallopian tube

Sperm transport within the tube appears to be a discontinuous process. It has been found that spermatozoa in the isthmic region just above the uterotubal junction remain there for several hours in the preovulatory

Fig. 5.6. Section of the isthmus of a greater horseshoe bat *Rhinolophus ferrumequinum* collected during hibernation; numerous spermatozoa line the surface of the mucous membrane. (From photomicrograph by C. R. Austin.)

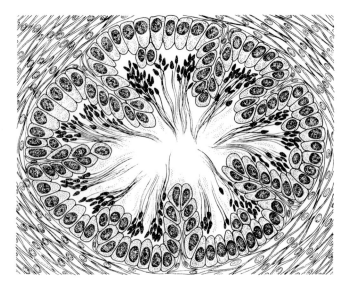

period, only moving up to the ampullary-isthmic junction (the site of fertilization) about the time of ovulation (Fig. 5.8). This behaviour has been documented in the mouse, pig and rabbit and may be more widespread. The reason for the temporary halt in the isthmus is not known, but in the upper regions of the tubes spermatozoa are much more active and have vigorously lashing tails, and this activity is thought to be important also in aiding sperm penetration through the investments surrounding the egg (see Chapter 6). In addition, muscular activity in both circular and longitudinal layers of the tube, and in the mesosalpinx, may be involved in moving spermatozoa from the isthmus to the site of fertilization, and in ensuring adequate mixing of tubal contents so that the chances of a spermatozoon meeting an egg are enhanced.

The epithelium of the Fallopian tube contains many secretory cells, especially in the isthmus, and these secretions fill the lumen of the tube, which is usually narrow because of the large and often branched mucosal folds. Except at oestrus or when the eggs are passing into the uterus following their period of tubal transport, the uterotubal junction (as mentioned above) is closed. Thus most of the tubal secretion flows passively into the peritoneal cavity. In those regions where there are many cilia, some fluid flow may be towards the uterus because most of the cilia

Fig. 5.7. Scanning electron micrograph of the uterotubal junction in a pig at oestrus, showing the complexity of the duct lumen and the oedematous protrusions of mucous membrane. (From R. H. F. Hunter. *Physiology and Technology of Reproduction in Female Domestic Animals*, Fig. IV.11. Academic Press; London. (1980).)

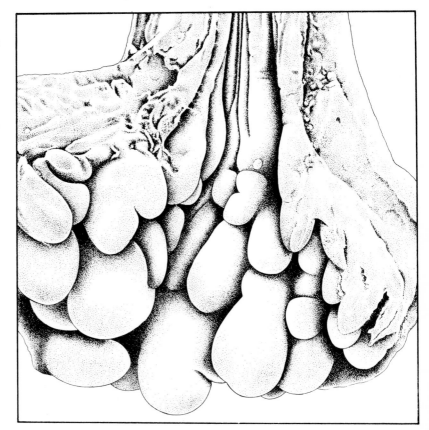

beat away from the ovary and towards the uterus. However, such a fluid flow is probably confined to a shallow layer over the epithelial surface, and, when the fluid reaches the constricted region of the tube at the ampullary-isthmic junction, it will reverse direction and flow back as an axial stream towards the open fimbriated end of the tube and thus into the peritoneal cavity.

The egg at ovulation is surrounded by several layers of follicle cells, radially arranged near the egg surface (see Chapter 2); this arrangement of cells may orient the sperm head towards the egg. People have suggested that the egg may itself release chemical messengers to attract spermatozoa

Fig. 5.8. Fallopian tubes of three rodents, rabbit and man, showing differences in structure. Broken lines indicate outlines of the intramural part of the tube. The arrowheads point approximately to the junction between isthmus and ampulla. (From O. Nilsson and S. Reinius. In *The Mammalian Oviduct*, ed. E. S. E. Hafez and R. J. Blandau. University of Chicago Press (1969).)

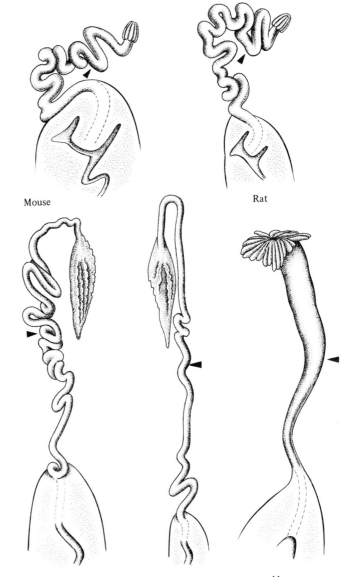

Mouse

Rat

Guinea pig

Rabbit

Man

and thus increase the chances of fertilization. Such a mechanism is known to exist in certain simple plants and animals, but there is no evidence for its existence in mammals.

Motile spermatozoa have been found in the human Fallopian tube up to 85 hours after coitus. However, fertilizing capacity is usually lost before motility and in most animals is limited to a day or two; estimates for the fertile life of spermatozoa are presented in Table 5.3. (Figures for the fertile life of eggs are set out in Table 3.2.)

In the species that have been studied, many of the spermatozoa reaching the ampullary portion of the Fallopian tube (see Figs. 5.8 to 5.11) are propelled into the peritoneal cavity – by their own motility, through the contractile activity of the smooth muscle in the wall of the tube, or by the flow of tubal secretion into the peritoneal cavity. Spermatozoa can remain alive in the peritoneal cavity for at least 24 hours and fertilization and early development have been initiated in the dog, fowl, rabbit, cow and guinea pig following intraperitoneal injection of a sperm suspension.

## Egg transport
### Structure of the Fallopian tube
For a better understanding of the process of entry of the egg into and movement along the Fallopian tube to the uterus, a brief description of the basic structure of the tube might be helpful. The Fallopian tube (also known as the oviduct and, in medical texts, the uterine tube) is a muscular tubular organ lined with secretory and ciliated epithelial cells, with four distinct anatomical regions (Fig. 5.9). The infundibulum is the trumpet-shaped, thin-walled portion adjacent to the ovary and it opens into the peritoneal cavity. This opening (the ostium) is surrounded by densely ciliated (Fig. 5.10), petal-like structures forming the fimbria (the 'fringe').

Table 5.3. *Estimates of the fertile life of spermatozoa in different animals*

| Animal | Hours |
| --- | --- |
| Mouse | 6 |
| Rat | 14 |
| Guinea pig | 21–22 |
| Rabbit | 30–32 |
| Ferret | 36–126 |
| Sheep | 30–48 |
| Cow | 28–50 |
| Horse | 144 |
| Man | 28–48 |
| Bat | 135 (days) |

*Source*: From B. J. Restall (1967). *Adv. Reprod. Physiol.* **2**, 181.

The fimbria in the human and rabbit is partially attached to the ovary, thus ensuring close proximity of the ostium to the ovarian surface.

The next segment of the tube is the ampulla. In humans this is the longest region (about 5–8 cm) and even in the rabbit it comprises at least 50 per cent of the length of the tube (about 10 cm). The ampullary lumen is quite large (1 cm near the ostium in the human) but is filled with branched mucosal folds (Fig. 5.11), and near its junction with the isthmus may be only 1–2 mm in diameter. The wall of the ampulla is relatively thin and comprised mainly of circular muscle with only a very small number of longitudinal fibres.

The third region of the tube, the isthmus, runs from the ampullary-isthmic junction to the uterotubal junction, and is about 2–3 cm in length in humans and some 8–10 cm long in rabbits. The wall has a very thick layer of circular muscle fibres and a thin one of longitudinal fibres. The lumen

Fig. 5.9. Diagram of the human ovary, Fallopian tube and part of the uterus to illustrate the relations between these structures and the egg and embryo at different stages of development. Broken lines (a), (b) and (c) show approximate positions of the sections portrayed in Fig. 5.11.

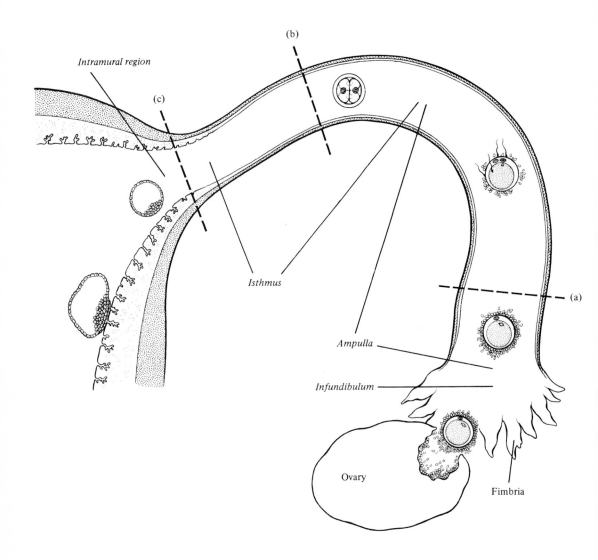

Fig. 5.10. A scanning
electron micrograph of the
cilia in the Fallopian tube,
illustrating the remarkable
density of these structures
at the ovarian end. The
relatively smooth cells are
probably mucous-secretory.
(From E. S. E. Hafez
(ed.). *Scanning Electron
Microscopic Atlas of
Mammalian Reproduction.*
Springer-Verlag; New
York. (1975).)

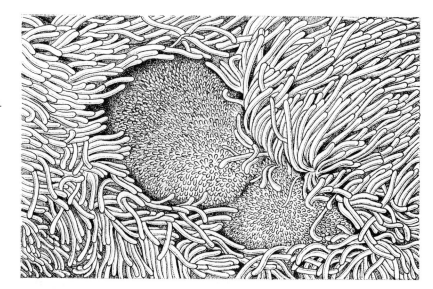

Fig. 5.11. Sections of the Fallopian tube in the regions of the ampulla (a), the isthmus (b) and the intramural part of the tube (c) (see Fig. 5.9), showing the large differences in lumen size and complexity of the luminal surface. (From W. Bloom and D. W. Fawcett. *A Textbook of Histology.* Saunders; Philadelphia (1969).)

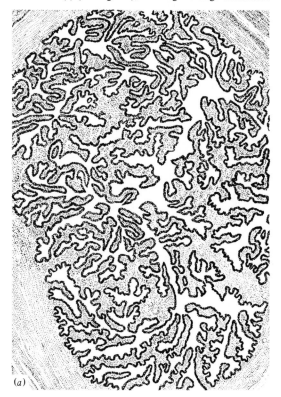

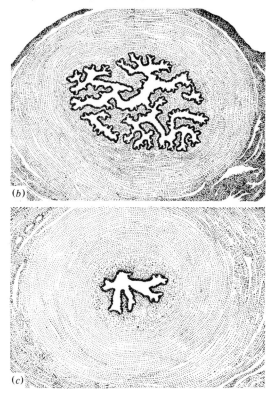

of the isthmus is narrow (about 0.4 mm) (Fig. 5.11). The epithelium lining the lumen is comprised mainly of secretory cells with only a few ciliated cells. The muscular wall of the isthmus is densely innervated with $\alpha$-adrenergic neurons, but their function in controlling egg transport is obscure since chemical denervation is without effect on this process. It is thought that the whole or a portion of the isthmus may act in the manner of a sphincter to retain eggs in the tube until the appropriate moment for their entry to the uterus.

The fourth and final portion of the Fallopian tube is the so-called intramural region because it lies within the uterine wall (Figs. 5.8 and 5.9). This section is much more developed in some species, such as primates, than in others, such as rodents. The length of this segment depends on the shape and size of the uterus and the thickness of the uterine wall. In humans it can be 1–2 cm long with a diameter of 0.1–1 mm. The uterotubal junction generally lies at the junction of the isthmus and the intramural segment and is much better defined in some species than in others.

Fig. 5.12. Graph showing how the mean amplitude of tubal contractions was increased following the injection of 100 IU hCG. The dotted line is the recording for the left tube of one rabbit and the other two lines for the left and right tubes of another rabbit. (From M. Salomy and M. J K. Harper. *Biol. Reprod.* **4**, 185 (1971).)

### Egg '*pick-up*'

There have been several theories to explain how eggs enter the Fallopian tube. The tube cannot exert any suction, but the vigorous action of the cilia on the fimbria is sufficient to create a fluid flow towards the tubal ostium, and the muscular activity of the tube and even of the ovary itself all seem to contribute to this. Movements of these various structures are increased at the time of ovulation (tubal contractions are greatly augmented

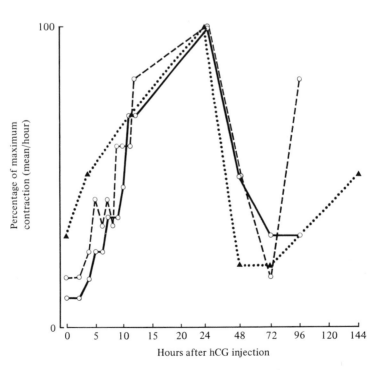

following injection of hCG – Fig. 5.12; the effect is presumably due to raised circulating levels of ovarian hormones). Thus, in those species without an ovarian bursa, both cilia and muscular activity probably play a major role in 'pick-up' or the recovery of the ovulated egg from the ovarian surface. In rodents, in which the ovary is almost totally enclosed by a fluid-filled bursa, fluid movements produced by cilia may be more important. In primates, the mechanism of 'pick-up' is probably like that in other species without a bursa, with the fimbria being brought into close association with the ovarian surface at ovulation.

Entry of eggs into the tube is a highly efficient process, and even trans-peritoneal migration of eggs – from one ovary to the contra-lateral tube – is said to be not uncommon in women. 'Pick-up' by the tube of eggs and other inert substances from the peritoneal cavity has been demonstrated in rabbits, and as we have already mentioned, spermatozoa injected into the peritoneal cavity have been shown to initiate pregnancy in several species. Attempts to prevent pregnancy by covering the fimbriated end of the tube with a plastic hood have not proved effective in monkeys, since eggs can apparently enter the tube through even very small openings. Egg 'pick-up' is evidently facilitated by the presence of the cumulus oophorus (see Chapter 3) surrounding the egg at the moment of ovulation, since this provides a larger surface on which the cilia can obtain purchase.

*Movement of the egg along the Fallopian tube*
When the egg is shed from the follicle it is surrounded by both the cumulus oophorus and corona radiata (Chapter 2). The cumulus is critically important for the normal transport of the egg through the ampullary

Fig. 5.13. Recordings showing the responses to certain agonists of longitudinal (LI) and circular (CI) muscle tissue preparations from the isthmus of the Fallopian tube, and the longitudinal (LA) and circular (CA) muscle tissue from the ampulla. The tubes were taken from a baboon one day before ovulation, and set up in a bath of Krebs's solution. The agents were in contact with the tissues for one minute at a final concentration of 1 $\mu$g/ml (free base). INE, isoprenaline; NE, noradrenaline; PGE$_2$, prostaglandin E$_2$; U44069, (15S)-hydroxy-9$\alpha$, 11$\alpha$-(epoxy-methano)prosta-5Z, 13E-dienoic acid; H, histamine; ACh, acetylcholine. (From A. Johns and L. W. Coons. *Biol. Reprod.* **25**, 120 (1981).)

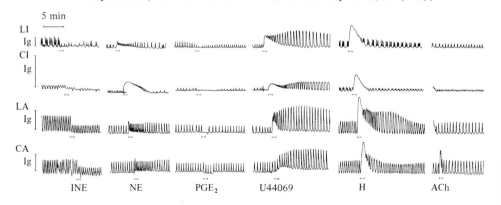

region in eutherian mammals, for which cilia are primarily responsible through their interaction with the surface of this cell mass. Surgical reversal of a segment of ampulla – so that the cilia beat in the reverse direction to normal, i.e. to the ovary from the uterus – prevents normal transport and causes the eggs to clump on the ovarian side of the reversed segment. Removal of cumulus and corona cells from eggs and paralysis of the cilia both interfere with normal transport through the ampulla. On the other hand inhibition of muscular contractility does not. Activity of the circular muscle of the ampulla though vigorous (Fig. 5.13) is not peristaltic and, in the absence of ciliary activity or of cumulus cells, the eggs may never reach the ampullary–isthmic junction, or if they do it is as a random event. However, we must record that women who suffer from the Immotile Cilia syndrome (in which cilia are non-functional) may have normal fertility. Perhaps they have some cilia still working, or else the role of cilia in egg transport may be more critical in some species, like the rabbit, than in man.

After the eggs reach the site of fertilization, the outer layer of cells is usually rapidly lost, partly through the mechanical action of cilia and

Table 5.4. *Time of egg denudation, entry into uterus and cell stage at entry, in relation to ovulation*

| Animal | Denudation of eggs after ovulation (hours) | Entry into uterus after ovulation (hours) | Cell stage at entry |
|---|---|---|---|
| Opossum | — | 24 | Pronuclear |
| Shrew | — | 87–95 | Late morula |
| Rat | Rapid | 95–100 | 8–16 |
| Mouse | 12 | 72 | 16–32 |
| Guinea pig | 24+ | 80–85 | 8 |
| Rabbit | 6 8–10[a] | 60 | Morula |
| Dog | 48 or more | 168–192 | 16–32 |
| Ferret | $19\frac{1}{2}$–40 | $88\frac{1}{2}$–$108\frac{1}{2}$ | 16–32 |
| Cat | 48 or more | 121–$161\frac{1}{2}$ | Late morula |
| Horse | < 24 | 98+[b] | 16 |
| Pig | < 24 | 24–48 | 4 |
| Cow | 9–14 | 72 | 8–16 |
| Goat | < $30\frac{1}{2}$ | 85–98 | 12 |
| Sheep | 0–5 | 70–80 | 8–16 |
| Rhesus monkey | 24 48[a] | 72±12 | 16 |
| Baboon | 48[a] | 72 | — |
| Man | ?24 | 80 | — |

Data from various sources.
[a] Unmated; [b] except unfertilized eggs.

muscular movements of the tube, and partly (in mated or artificially inseminated animals) through the presence of hyaluronidase, an enzyme released from the sperm acrosome.

In most species fertilization seems to take place in the ampulla, or at the ampullary–isthmic junction in species in which transport is very rapid, like the rabbit. Although egg transport through the ampullary region of the oviduct takes only a few minutes in the rabbit, in other species, such as the sheep, the human and the rhesus monkey, it can take over 24 hours. In species with rapid transport, the eggs are normally retained at the ampullary–isthmic junction for up to 24 hours after ovulation. Then slow transport through the isthmus takes place. Despite these variations, the overall time course of egg transport through the entire length of the tube is remarkably similar, taking approximately 3 days in most species (Table 5.4). Some exceptions have been observed, such as the opossum and the pig, in which transport is faster, and carnivores, in which it is slower. In most species the transport time is the same irrespective of whether the eggs are fertilized or not, but this is not true for horses and donkeys where an unfertilized egg may remain in the tube for weeks or months and actually be by-passed by a subsequently ovulated and fertilized egg. The Equidae appear to be unique in this respect.

Parenthetically, we should note that if fertilization has occurred, what we are still calling an egg is in fact an embryo, but it is convenient and indeed common practice to use the former term, at least until entry into the uterus. (The life of the embryo – and fetus – is the subject of Chapters 1 and 2 in Book 2.)

We believe that the main method by which eggs are transported through the isthmic region of the tube is by contractions of the circular smooth-muscle layer. Ciliary or fluid forces seem unlikely to exert any major action in this portion of the tube. Several agents increase or decrease the activity of tubal musculature (Fig. 5.13), but these do not appear to influence consistently the speed of egg transport. The characteristic muscular activity of the isthmus does not propagate in the same way as the intestine, that is, it is not peristaltic. Contraction seems to be random, and propagation of electrical activity takes place over only short distances imparting a to-and-fro movement to the egg. Over the 3-day period of transport, a net directional bias towards the uterus is thought to develop, and this finally results in movement of the egg into the uterus. Associated with this more directed muscular activity there is a relaxation of the circular muscle layer of the isthmus which increases the luminal diameter and so permits egg passage. It is rare to find eggs in the last 10 per cent of the isthmus just above the uterotubal junction, even just prior to their entry into the uterus; it is common to find them clumped just above this region. This suggests that final passage through the region is very rapid and depends essentially on relaxation of the muscle wall. At this time any fluid collected in the Fallopian tube will also rapidly pass into the uterus.

*The egg in the uterus*

Once eggs (now morulae or blastocysts) have passed into the uterus they tend to remain at the tubal end of the uterine lumen in non-primates for a period of time before muscular contractions space them equally along the length of the uterine horn. The exact mechanism whereby such even spacing is arranged in species in which multiple embryos implant simultaneously is not known. One suggestion is that a blastocyst may exert a local stimulatory action on the uterine smooth muscle (the myometrium), causing contractions to be propagated in both directions away from the position of each blastocyst. These contractions would become weaker as they got further from the point of origin, and at some point would be equalized by contractions propagating in the opposite direction from another blastocyst. In this manner the equalizing forces could gradually ensure that the blastocysts would be evenly spaced along the length of the uterine lumen. Another possible explanation for the spacing is that regions of the endometrium around an implanting blastocyst become refractory to other blastocysts. These possibilities are discussed further in Book 2, Chapter 2.

At the time that these things are happening, the uterus is under the influence of progesterone produced by the corpus luteum. This ensures that myometrial activity is fairly quiescent and prevents expulsion of the developing embryos through the cervix into the hostile environment of the vagina. In primates, where there is only a single uterine cavity and generally only one egg is ovulated during each cycle, spacing of blastocysts in the uterus is not a problem. Even in primates, however, luteal phase progesterone keeps myometrial activity minimal.

*Conditions influencing egg transport*

The exact controlling mechanisms that determine the programmed transport of the egg to the uterus at the appropriate time are still obscure. Ovarian hormones, oestradiol and progesterone, can exert a major influence on the activity of the tubal musculature and on the rate of transport. In many species, depending on the dose given, oestrogens can cause either retention of eggs in the tube beyond the normal time ('tubal-locking') or accelerated transport to the uterus. Oestrogen seems to be more effective in causing tubal locking if the dose is supraphysiological and is given at or just before ovulation, when the eggs have still not passed the ampullary–isthmic junction. Once the eggs have reached the isthmus, regardless of dose, rapid transport to the uterus ensues. An action of oestrogen on tubal transport of eggs could be the major mechanism whereby emergency treatment with very large doses of oestrogens after isolated mid-cycle intercourse provides effective contraception in women. In this case the action is probably by accelerated transport through the tube and out of the uterus, rather than reversal of transport with passage of the egg into the peritoneal cavity, because there is no increase in the 'ectopic' pregnancy rate. (Ectopic pregnancy is discussed in the next section.) The treatment can still be successful if started up to 72 hours after intercourse, long after 'tubal-locking' might be expected to occur. Post-coital oestrogen treatment is also an effective way of preventing pregnancy in bitches after a mis-mating.

By contrast, the administration of progesterone for 3 days prior to ovulation causes accelerated transport in the rabbit, but treatment with a similar regimen after ovulation is ineffective. We do not know whether a similar effect occurs in primates, but can infer that it is unlikely from the not infrequent occurrence of pregnancies in women taking progestin-only contraceptive pills.

How these hormones exert these effects is still unknown. Whether the action is directly on the smooth-muscle cells and their membrane potential, or via induction of protein synthesis, or indirectly through changes in prostaglandin, noradrenaline or cyclic nucleotide levels, has still to be established. Inhibition of adrenergic activity chemically by 6-hydroxy-dopamine, or surgically, or by depletion of peripheral adrenergic stores with reserpine, has little effect on egg transport. Similarly, inhibition of prostaglandin synthesis with indomethacin is ineffective. It may be, of course, that all the control systems overlap, so that if one is inhibited, another might take over. For the moment all this is speculation, although it is is known that there are specific receptors for different prostaglandins (PG), such as $PGE_2$ or $PGF_{2\alpha}$, on tubal smooth-muscle cells, and that binding of these prostaglandins is associated with changes in calcium uptake. Calcium is necessary for the contraction of smooth-muscle cells. Furthermore the number of these prostaglandin receptors changes in different segments of the tube as egg transport is occurring. We do not yet know whether such changes are directly related to mechanisms controlling transport, but they are likely to be associated with the contraction or relaxation of individual smooth-muscle cells.

*Ectopic pregnancy*
In the foregoing discussion on egg transport the term 'ectopic' pregnancy was used. This means the implantation and development of a fertilized egg in a location other than the uterus (usually in the Fallopian tube but sometimes in the ovary or in the peritoneal cavity). Abnormal implantations can be produced experimentally in the abdominal cavity, in the eye, in the kidney and even in the testis of laboratory rodents, as David Kirby demonstrated, but such developing embryos generally die at a relatively early stage (see Book 2, Chapter 2). By contrast, intra-abdominal, intra-ovarian and intra-tubal pregnancies occur naturally in women without any sort of experimental intervention. Such occurrences are not infrequent, and are often the result of some tubal pathology caused by diseases, such as tuberculosis or venereal disease. Undiagnosed ectopic pregnancies lying outside the tube can progress to full term and beyond,

but because of size constraints those within the tube soon cause a rupture of the tubal wall with associated pain, severe haemorrhage and shock. Accordingly they have to be removed surgically, preferably before rupture, to save the life of the mother.

Naturally occurring ectopic pregnancies seem to be a peculiarity of the human, because very few have been recorded in other primates and none in non-primates. Nor has anybody been able to produce intratubal implantations experimentally in non-primates; eggs (embryos) that remain in the tube longer than usual die rapidly, either because of deleterious conditions in the tube or more likely because of the absence of some uterine factor. Tubal eggs at early stages of development have rather special needs and do not survive well if transferred precociously to the uterus in non-primates; in monkeys, though, transfer of newly fertilized tubal eggs to the uterus can give rise to normal pregnancies. In humans, there is a rare procedure known as Estes's Operation in which the ovary is surgically inserted through the uterine wall so that it lies in the uterine lumen and ovulation occurs directly into the uterine cavity. There are claims (often disputed) that in a small percentage of such cases the patient achieved a normal pregnancy. Thus, in primates the requirement for the orderly progression of tubal transport and embryonic development prior to uterine entry seems much less critical than in non-primates, emphasizing the dangers of drawing analogies between different species.

Gamete transport is the vital link between gamete production and syngamy, and it involves very many events and regulatory mechanisms. This is an attractive area for research where we might be able to discover new ways to regulate fertility. Physical barriers to gamete transport, such as vasectomy, tubal ligation, the condom and the diaphragm are still among the safest and most effective forms of contraception, but we may yet be able to improve upon them. And with the growing interest in culture and fertilization *in vitro*, we are becoming increasingly conscious that there is a great deal more to be learned about the situation *in vivo* that we are trying to emulate.

### Suggested further reading

Fertilization in relation to the number of spermatozoa in the Fallopian tubes of rabbits. M. C. Chang. *Annal. Ostet. Ginec.* **73**, 918–25 (1951).

Reaction of the uterus on spermatozoa in the rabbit. M. C. Chang. *Annal. Ostet. Ginec.* **78**, 74–86 (1956).

Studies on the duration of ovum transport by the human oviduct. IV. Time interval between the luteinizing hormone peak and recovery of ova by transcervical flushing of the uterus in normal women. S. Diaz, M. E. Ortiz and H. B. Croxatto. *Amer. J. Obstet. Gynec.* **137**, 116–21 (1980).

Detailed time course of ovum transport in the Rhesus monkey (*Macaca mulatta*). C. A. Eddy, R. G. Garcia, D. C. Kraemer and C. J. Pauerstein. *Biol. Reprod.* **13**, 363–9 (1975).

Pattern and duration of ovum transport in the baboon (*Papio anubis*).
C. A. Eddy, T. T. Turner, D. C. Kraemer and C. J. Pauerstein. *Obstet. Gynec.* **47**, 658–64 (1976).

Function and malfunction of the Fallopian tubes in relation to gametes, embryos and hormones. R. H. F. Hunter. *Eur. J. Obstet. Gynec. Reprod. Biol.* **7**, 267–83 (1977).

Concentrations of serum elements in the intraluminal fluids of the rat seminiferous tubules, rete testis and epididymis. A. D. Jenkins, C. P. Lechene and S. S. Howards. *Biol. Reprod.* **23**, 981–7 (1980).

Maturation, transport and fate of spermatozoa in the epididymis.
J. M. Bedford. In *Handbook of Physiology, Section 7: Endocrinology, vol. 5 Male Reproductive System.* Ed. D. W. Hamilton and R. O. Greep, pp. 303–17. American Physiological Society, Washington, DC (1975).

Gamete transport, fertilization and implantation. C. A. Eddy and M. J. K. Harper. In *Fertility Control. Biologic and behavioral aspects.* Ed. R. N. Shain and C. J. Pauerstein, pp. 32–48. Harper and Row; Hagerstown, Md. (1980).

Parameters of male fertility. R. Eliasson. In *Human Reproduction. Conception and Contraception.* Ed. E. S. E. Hafez and T. N. Evans, pp. 39–51. Harper and Row; Hagerstown, Md. (1973).

Mammalian sperm movement in the secretion of the male and female genital tracts. D. F. Katz and J. W. Overstreet. In *Testicular Development, Structure and Function.* Ed. A. Steinberger and E. Steinberger, pp. 481–9. Raven Press; New York (1980).

*Ovum transport and fertility regulation.* Ed. M. J. K. Harper, C. J. Pauerstein, C. E. Adams, E. M. Coutinho, H. B. Croxatto and D. M. Paton, 566 pp. Scriptor; Copenhagen (1976).

# 6

# Fertilization

*J. MICHAEL BEDFORD*

Biologists see fertilization as an intriguing event, at least partly because it represents a complex interaction between unusually specialized cell types, and they pursue the study of fertilization now with the tools of modern cell biology and of biochemistry. Fertilization has a fascination also for a wider

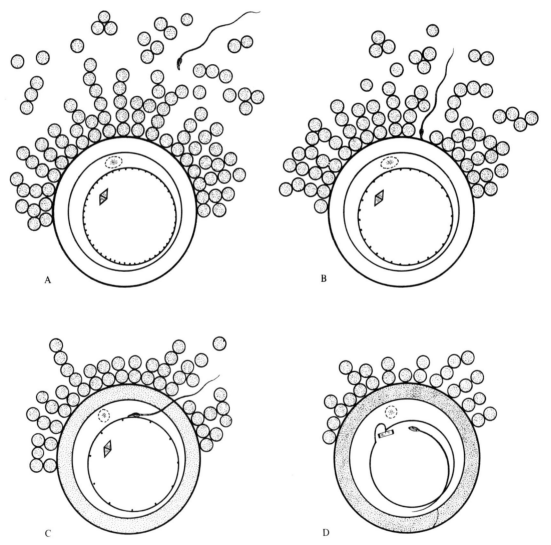

audience. This stems from fertilization being the point at which the *potential* for new life appears; and from a practical standpoint its study is important to everyone because it is a phase of conception that is relatively untapped in the worlds of contraceptive development and infertility. Fertilization is of course not just one event. It comprises a series of steps beginning with penetration of the egg coats followed by incorporation of the spermatozoon into the cytoplasm of the egg, and then the activation of the egg. Next there occurs a transformation of the sperm nucleus and of the remaining haploid set of egg chromosomes so as to form, respectively, the male and female pronuclei. The approach of the pronuclei

Fig. 6.1. Stages of fertilization of the rat egg. A. Initial contact of a spermatozoon with the cumulus oophorus soon after ovulation. B. Binding of spermatozoon to the surface of the zona pellucida after it has passed through the cumulus oophorus. C. Moment of sperm attachment and fusion with the oolemma after penetration of zona pellucida. This is accompanied by some contraction of the vitellus and so enlargement of the perivitelline space. (Shading of zona depicts the beginnings of the zona reaction induced by the content of the erupting cortical granules.) D. The sperm head begins to swell and emission of the second polar body is under way (note slit in zona left by sperm entry). E. Emission of the second polar body is complete and the sperm chromatin is decondensed. F, G. Both pronuclei have formed, the cells of the cumulus oophorus have almost dispersed and in G the two pronuclei have come into apposition. H. The chromosomes of the male and female are aligned on a mitotic spindle whose progress from metaphase (shown here) through to telophase heralds the first cleavage division of the new embryo. (Redrawn from Text-fig. 1 in C. R. Austin and M. W. H. Bishop, *Biol. Rev.* **32**, 296 (1957).)

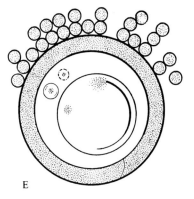

E

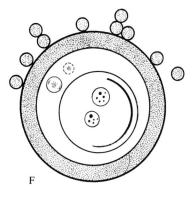

F

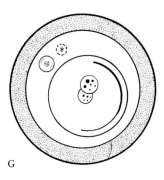

G

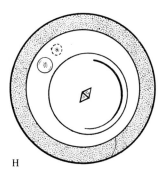

H

and the eventual pairing of the chromosomes that later differentiate within each pronucleus constitute syngamy – the final step which occurs 12 hours or more after the spermatozoon first enters the egg. This general sequence is shown in Fig. 6.1.

A variety of other elements make fertilization in mammals a more complex matter in its biological setting. Until very recently it was believed that mammalian fertilization closely resembles that described for simple creatures, other than the fact that the mammalian process is internal and occurs within the confines of the ampulla of the Fallopian tube. We now know that in many respects this is not so, and that the physiological relationships of mammalian gametes with their environment and the mode of fertilization itself differ in important details from those established for the sea urchin and other invertebrates in the classical studies of Frank Lillie, Jacques Loeb and others. Not only do mammalian spermatozoa need to undergo a complex phase of maturation over a period of several days after they leave the testis, but they must also complete what can be thought of as an extra step of maturation, which occurs in the female tract, namely capacitation. Very few mammalian spermatozoa actually reach the ampulla of the oviduct where fertilization occurs, and because so few are there one might expect that the egg could attract spermatozoa to it – a process known as chemotaxis, which certainly exists in some simple animals such as marine hydroids. However, for mammals there is no evidence of such a system of communication between the spermatozoon and the egg, and their eventual contact may well be a matter of chance. It may be that mammalian gametes have dispensed with another mode of communication, used by sea urchins and a variety of other groups, in which surface-active molecules from the egg coat are able to induce a controlled disruption of the sperm's acrosome. In addition, the particular way the spermatozoon actually enters into or is incorporated by the egg of eutherian mammals is very different from that in other animals. We must bear in mind, too, that as in other aspects of reproduction elements of the fertilization process vary even among different mammals: a fact for one species may not be so for others.

Many of the preliminaries to fertilization and details of the structure of the gametes themselves have been described in earlier chapters. We can consider now specifically the immediate preparation that each must undergo to be able to participate in gamete union, the sequential steps and postulated mechanisms involved in that process, how it may sometimes result in anomalies that cause embryonic failure, and the recently developed manipulations that are beginning to bring the understanding of fertilization directly into the clinical realm.

**Final preparation of gametes**

*Oocytes*

The period between the 'surge' of serum levels of luteinizing hormone and the ovulation it stimulates brings changes both in the follicle and the oocyte. Primary oocytes with a germinal vesicle, the arrested diplotene or dictyotene nucleus, cannot be fertilized normally, and critical changes must occur then that make them competent to support normal development when penetrated by a spermatozoon. As explained by Terry Baker in Chapter 2, the hormonal stimulus for ovulation also triggers the resumption of meiosis, such that in the hours preceding ovulation the germinal vesicle breaks down, the first polar body is shed and meiosis proceeds as far as metaphase II. There it halts until stimulated further to its completion, usually by entry of the fertilizing spermatozoon. But exactly how does this limited resumption of meiosis affect the fertilizability of the oocyte? Although the cumulus oophorus of the oocyte from the unstimulated follicle is distinctly less susceptible *in vitro* to dispersion by hydrolytic enzymes, including those from the sperm acrosome, this cell mass in its immature state (as seen around primary oocytes in hamster, guinea pig and rabbit, at least) does not seem to be a notable barrier to spermatozoa in the oviduct. Neither in the mammals examined carefully is the zona pellucida any more of a barrier in the primary oocyte than in the ovulated oocyte. Finally, even the plasma membrane of the primary oocyte readily allows a penetrating spermatozoon to fuse with it. The critical changes that confer normal fertilizability are reflected chiefly, therefore, in a developing capacity of the oocyte to take part in one or more of four events: sperm-head engulfment, exocytosis of cortical granules, dispersal of sperm nuclear envelope, and decondensation of the sperm chromatin (Fig. 6.2).

Though not necessarily true in all mammalian species (for example, note the case of the hamster in Fig. 6.12), Miguel Berrios and I found that immaturity of the rabbit primary oocyte is expressed often in a poor ability of the egg cortex to rise up around and engulf the front of the sperm head as it does in ovulated eggs (compare A and D in Fig. 6.2), and many cortical granules at the extreme periphery of the oocyte fail to undergo exocytosis in response to the fusion stimulus. In addition to these aspects that could result from an immaturity of (cytoskeletal?) elements of the oocyte cortex or a failure of calcium ion influx, there is always an obvious inability of the immature ooplasm to deal with the sperm head. Charles Thibault and Micheline Gerard first drew attention to its striking lack of the power to transform the penetrated sperm head into a pronucleus. Ultrastructural studies suggest that this inability can result from lack of a response toward the sperm nuclear material as such, or also, in the case of the rabbit at least, persistence of the sperm nuclear envelope. This disappears immediately within mature ooplasm, but remains for hours around the sperm head that penetrates the primary oocyte. Maturation of these functions seems to begin with breakdown of the germinal vesicle, and it is possible that this releases activating 'factors' in the ooplasm.

An exception in what otherwise seems a general principle is provided by the dog oocyte, for this is ovulated and is able to support limited male pronucleus development while its germinal vesicle is still intact.

Fig. 6.2. Comparison of the responses to a fertilizing spermatozoon shown by a mature ovulated oocyte (A–C, left) and an immature primary oocyte (D–F, right). The deficiencies of the follicular egg are expressed in the absence or minimal character of the engulfment of the inner acrosome membrane, failure of dispersal of the sperm nuclear envelope and the greatly retarded decondensation of the sperm chromatin. In E and F, 'e?' is thought to be the equatorial segment which often seems to become separated in the immature rabbit oocyte. The persistence of cortical granules in the follicular oocyte is correlated with the common occurrence of polyspermy. Rabbit gametes. (Fig. 1 in M. Berrios and J. M. Bedford, *J. Cell Sci.* **39**, 1–12 (1979).)

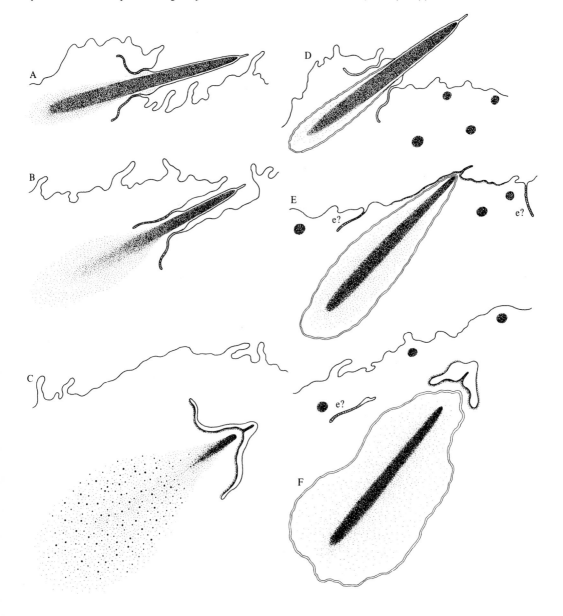

*Spermatozoa*

As explained in detail by Brian Setchell in Chapter 4 and Mike Harper in Chapter 5, spermatozoa are formed by a process that involves initial proliferation of primitive germ cells through cell division, and then a complex differentiation that gives each cell the characteristic appearance of the spermatozoon. But even when released from the testis, spermatozoa are not competent to swim properly or to fertilize. They must experience a phase of further maturation as free cells over the succeeding few days as they pass through the upper part of the epididymis. If we look back to earlier steps in vertebrate evolution, however, we find that spermatozoa released from the testis of, for instance, teleost fish or the frog will fertilize immediately and need no phase of epididymal maturation to be able to do so. What then is the biological significance of the epididymal maturation? The beginning of a sperm maturation function for the Wolffian duct is evident in many sub-therian vertebrates that have adopted internal fertilization, but solely as a maturation of their capacity for independent motility. This is so in elasmobranch fish, some reptiles, the chicken and other birds such as ducks and pigeons, and in the echidna, a monotreme mammal.

Then there seems to have been an almost quantum leap in complexity of epididymal function with the evolutionary emergence of marsupials and eutherian mammals, in which maturation of fertilizing ability involves several additional changes in the sperm cell. Why this further entrainment of the Wolffian duct for sperm maturation should have occurred is not quite clear still. One explanation rests on the possibility that the modifications spermatozoa undergo in the epididymis may have arisen in response to conditions imposed within the female tract and, in mammals, additionally perhaps because of evolutionary change in the character of the egg itself. Since we are not able properly to understand the biological significance of epididymal maturation, a variety of observations have been made essentially by chance, and even now many aspects are being studied intensely in something of a conceptual vacuum.

The epididymal maturation process in eutherian and marsupial mammals is reflected in most organelles in the sperm cell as it passes from the testis to the point of storage in the tail of the epididymis. One change that has a very obvious bearing on the ability to fertilize is a development of the ability for sustained progressive motility. We now believe that this directed movement is essential if the spermatozoon is to pass through the coats around the newly ovulated egg. This development is accompanied by metabolic change, reflected as one element in an accumulation by the spermatozoon of cyclic AMP, which in turn seems to bear upon the supply of energy to the axonemal elements of the tail. Another change is reflected in the molecular character of the sperm surface that in the guinea pig and some members of the squirrel family results very obviously in the stacking of sperm heads into rouleaux, and in American marsupials in a pairing of spermatozoa by the acrosomal face. In other species surface change in the maturing spermatozoa is only revealed more subtly by a variety of techniques. It seems attributable in part to acquisition of glycoproteins secreted by the epithelium of the epididymis, though such a single explanation may well be an oversimplification of what is really a whole spectrum of events that ensure the mature state of the surface of a functional spermatozoon. These surface changes probably have some bearing at least on the ability of the spermatozoon to bind to the zona pellucida.

Other changes of a structural nature increase the stability of the chromatin of the sperm nucleus. These result from the formation of strong disulphide (–S—S–) bonds between the cysteine-rich protamines that characterize, and are a major constituent of, the nucleus of the spermatozoa of eutherian mammals. A similar cross-linking of protein-bound -SH seems to occur in the dense fibres, mitochondrial membranes and sheath of the sperm tail, and could have some effect on the character of the swimming movement in mature spermatozoa. In all, such stabilization of head and tail means that the mature spermatozoon becomes a tough keratinoid cell.

Surprisingly perhaps, the modifications that occur in mammalian spermatozoa in the epididymis do not complete the process of maturation. Already 30 years ago, Bunny Austin and M. C. Chang both produced evidence that spermatozoa must also be exposed to the environment of the female tract for a few hours before reaching the functional state in which they may fertilize an egg. It is still not easy to understand the biological significance of the need for this additional step of 'capacitation' (but see the next section on the acrosome reaction), particularly because it is not clear yet exactly where in evolution the need has appeared among the vertebrates. Results I have obtained with John Rodger in one marsupial, the American opossum *Didelphis*, show that single progressively motile spermatozoa released from the epididymis or vas deferens will not penetrate eggs *in vitro*, whereas those flushed from the oviduct some hours after mating do so readily under the same conditions. Obviously this suggests for marsupials the need for capacitation that is seen in all eutherian mammals. Since there appears to have been a dramatic increase in the complexity of epididymal maturation with the evolution of marsupial and eutherian mammals, one must wonder whether the subsequent need for capacitation in the female is somehow linked functionally to the events of epididymal maturation in the male.

Nothing is known yet of what it is that capacitates spermatozoa in the female tract, but it does not seem to be a species-specific process. For example, the oviduct of a mouse will capacitate rabbit spermatozoa and allow them to fertilize rabbit eggs transferred to the mouse oviduct. However, for at least those species whose spermatozoa are deposited at coitus only as far as the vagina, capacitation appears to occur most rapidly if they pass through both uterus and oviduct, as they do naturally.

Moreover, since progesterone can inhibit the ability of the uterus to capacitate spermatozoa, it seems likely that distinct factors effect capacitation in the female tract.

It has been difficult to pinpoint precisely the cellular changes in the spermatozoon that are a function of capacitation, not least because the population recovered at any time from any region of the tract tends to be a mixture of spermatozoa in different states; some are dead and/or damaged, others are vigorously motile. Nonetheless, evidence coming from the behaviour of spermatozoa maintained under capacitating conditions in the female or *in vitro* suggests that an important component of it is a subtle though readily observable change in the beat of the sperm tail. First detected by Ryuzo Yanagimachi as a more vigorous and more undulating movement of the hamster sperm tail, a similar effect has been seen in other species since then. This urgent tail beat may develop 'in anticipation' of the imminent need to penetrate the barriers of the cumulus oophorus and especially the zona pellucida. Circumstantial evidence suggests that a second facet of capacitation may take the form of removal or modification of some protein components at the sperm surface. This seems essential as a pre-requisite for the acrosome reaction to occur in motile spermatozoa, and as suggested below could even regulate it.

*The acrosome reaction*

As described in Chapter 4, the anterior half of the sperm head carries the acrosome, which is moulded over the nucleus and is covered externally by the plasma membrane. A spermatozoon with an intact acrosome cannot penetrate an egg – for this, the acrosome must 'react' or break down, releasing its contents, and then the reacted elements must be shed. Successive stages of the acrosome reaction are presented diagrammatically in Fig. 6.3. First, the outer acrosome membrane and the plasma membrane of the sperm cell unite by fusion at a number of points over the front half of the sperm head. The spaces developing then provide exit ports for soluble acrosome contents, probably at least a dozen different enzymes in all, and in a short time the acrosome is emptied. The reacted shroud of fused vesiculating membranes is then shed in the process identified separately as acrosome *loss*, and this seems to occur an appreciable time after the acrosome *reaction*. The significance of the delay could be that the spaces (exit ports) between the vesicles enlarge quite slowly so that release of acrosomal enzymes is protracted, providing full opportunity for the lytic action of the enzymes to aid the passage of the spermatozoon through the egg investments. Loss of the acrosome and its visible content occurs before the sperm head makes serious inroads into the substance of the zona pellucida.

Of the dozen or so enzymes known to exist in the acrosome, the most important for sperm function in fertilization would appear to be hyaluronidase and a powerful trypsin-like enzyme called acrosin. As Stan

Meizel and his colleagues showed, acrosin is stored in the form of its zymogen pro-acrosin, which is converted into the active enzyme on its release in the acrosome reaction. Both enzymes could play important roles in sperm penetration of the egg coats but, as discussed later, exactly how they are deployed in the course of penetration is still rather obscure. For example, the matrix of the cumulus may well contain carbohydrates other than hyaluronic acid, and other acrosomal enzymes such as aryl sulphatase may act co-operatively in sperm penetration. Better understanding of this phase will come, we hope, with more critical investigations.

Fig. 6.3. Stages of the acrosome reaction in the eutherian spermatozoon, as seen in ultra-thin sections. (*a*) Intact acrosome in an unstimulated spermatozoon, showing the typical irregular contour of the plasma membrane over the acrosome. (*b*) As a first reaction response, multiple point fusions develop between the plasma membrane and outer acrosome membrane such that, in thin sections, these appear to constitute an array of vesicles. The acrosome content then swells and escapes through the ports that the fusion points create. This effectively completes the reaction. (*c*) The next step is acrosome loss which occurs in the fertilizing spermatozoon as it begins to penetrate the substance of the zona. AC, acrosome; OA, outer acrosome membrane; PM, plasma membrane; ES, equatorial segment whose limits are defined by arrowheads; IA, inner acrosome membrane. (After Figure 14 in J. M. Bedford and G. W. Cooper, *Cell Surface Reviews*, vol. 5, *Membrane Fusion*, pp. 66–125. Ed. G. Poste and G. L. Nicolson. Elsevier/North-Holland Biomedical Press; Amsterdam (1978).)

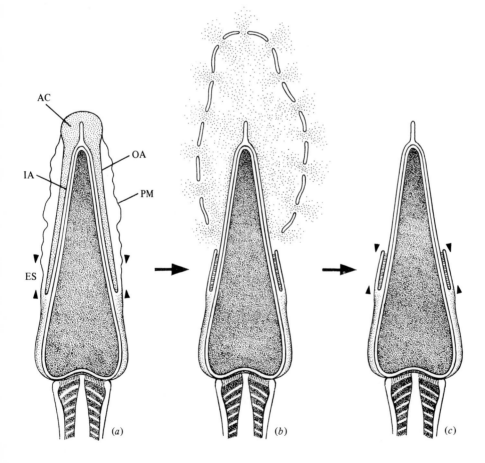

(*a*)          (*b*)          (*c*)

*Stimulus for the acrosome reaction.* As we have learned more about the details of fertilization, paradoxically there has come to be a greater uncertainty about the exact nature of the physiological stimulus that initiates the reaction in the oviduct. It has been tempting to speculate that the stimulus for the acrosome reaction in mammals might emanate from the egg coverings, as 'fertilizin' does so obviously in many marine invertebrates. Attempts to demonstrate this have not been successful and it is difficult to avoid the conclusion now that the controls for the acrosome reaction in mammals have changed radically in this respect in the course of evolution. There are several salient facts that speak against the possibility of a fertilizin in mammals. Spermatozoa react and then penetrate normally the egg devoid of granulosa cells and, *in vitro*, devoid even of a zona pellucida; they penetrate the zona pellucida of eggs treated in ways that could destroy the potency of a fertilizin-like activator, and of eggs that are aged or even already dead for one or two days. The possibility that the fertilization site might contribute a specific trigger is opposed by the fact that eggs can be fertilized in the uterus in some species, e.g. rabbit, as well as in a truly foreign tract (e.g. rabbit gametes in hamster, monkey in rabbit, cow in pig), and that for rodent spermatozoa in particular the acrosome reaction may occur *in vitro* in suitable media devoid of tract or egg components. One component, follicular fluid, fosters the reaction *in vitro*, but this fluid is not essential for it to occur nor for a normal rate of fertilization in the Fallopian tube when spermatozoa are transported in the normal way from the vagina. It is unlikely that the acrosome reaction is induced by the components of the serum complement sequence that damage cell membranes in conjunction with surface antigen/antibody reactions, since I have found that sperm penetration occurs consistently in the female rabbit depleted of complement for several days.

In summary then, various studies conducted both *in vivo* and *in vitro* have failed to provide evidence in Eutheria for the precedent established in lower animals that the acrosome reaction is stimulated by specific organic factors external to the spermatozoon. The interesting likelihood presents itself, therefore, that a new mechanism has arisen. Consideration should be given to the possibility that the reaction is regulated primarily by capacitation. As we shall see, calcium ions ($Ca^{2+}$) appear to be able to promote the acrosome reaction in capacitated spermatozoa. If the surface aspects of capacitation actually represent changes that allow $Ca^{2+}$ to move into the sperm head to promote membrane fusion, and if the sperm population expresses some heterogeneity of several hours duration in respect to the time at which different cells reach this point of capacitation, such an arrangement would avoid the need for a specific stimulus other than $Ca^{2+}$ in the environment. It would also allow for sequential small populations of spermatozoa to react spontaneously over a time that spans the pre- and post-ovulatory periods (Fig. 6.4). This idea needs a rather careful consideration of the dynamics of spermatozoa in the different compartments of the female tract, as well as the relative ability of each compartment to capacitate the spermatozoon. Whatever the outcome, this does at least seem to agree with all that has been observed experimentally, and is an hypothesis that could be tested.

*Mechanism of membrane fusion in the acrosome reaction.* This brings us to a second problem: the molecular mechanism of the fusion of plasma and acrosomal membranes. The literature of recent years contains a wealth of studies, largely *in vitro*, directed to this in one way or another. But it provides no unifying basis yet to explain the mechanism of the point fusions. Moreover, some of the findings tend to present a conflict of interpretation.

The work of Barry Bavister and Stan Meizel showed that agents in follicular fluid that promoted the acrosome reaction were those present also in very high concentration in the cortex and medulla of the adrenal gland. More specifically certain catecholamines, especially the $\beta$-adrenergic agonists, were active in this process and these are thought to function by stimulating membrane-bound adenylate cyclase, thus producing an increase in local levels of cyclic AMP. Cyclic nucleotides can provoke the acrosome reaction under certain circumstances, but the explanation as to why they can do so is elusive. Clearly more information is needed before we can claim to have a proper understanding of the energetics of the acrosome reaction.

In the cell systems examined in close detail, membrane fusions occurring at point locations appear to be preceded by a change in the distribution of the membrane particles that are made visible by freeze-fracture techniques. These particles exist in vast numbers and are probably aggregates of protein molecules; since they would carry relatively high electrostatic charges their presence is thought to be responsible, in part at least, for preventing membranes of cells from coming so close together that fusion would inevitably follow. If the distribution of the particles is changed in such a way as to clear them completely from small areas, these points are deprived of the repellent effect of the electrostatic charges and so would be free to make the close approximation leading to fusion. Extending the

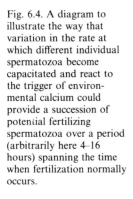

Fig. 6.4. A diagram to illustrate the way that variation in the rate at which different individual spermatozoa become capacitated and react to the trigger of environmental calcium could provide a succession of potential fertilizing spermatozoa over a period (arbitrarily here 4–16 hours) spanning the time when fertilization normally occurs.

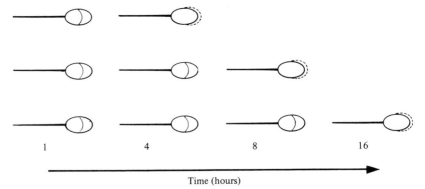

1                4                8                16

Time (hours)

idea to the mammalian acrosome, it is possible to identify small areas of both the outer acrosome membrane and overlying plasma membrane where particle density is low; and so we are tempted to infer that these are the future sites of fusion and that all we need look for in order to explain the mechanism of the acrosome reaction is the means whereby particle density can be reduced still further, to the level permitting close approximation and fusion. Such a change in particle distribution would presumably be energy dependent.

But whatever the precise events that lead to fusion, as noted above this depends on an influx of calcium. This ion is widely implicated in fusion events in other cell systems, and the most convincing evidence of its essential role for the acrosome reaction is two-fold. It comes first from the laboratory of Ryuzo Yanagimachi, where use was made of the guinea pig spermatozoon whose large acrosome can be assessed quantitatively in the light microscope; a progressive and ultrastructurally normal reaction was seen within a few minutes when calcium was added to the sperm incubation medium at successive intervals of time (Fig. 6.5). The reaction always failed to occur in motile spermatozoa in the absence of calcium. Subsequently, David Green likened the acrosome reaction directly to the stimulus–secretion coupling of exocytosis that calcium modulates in other cells, and almost certainly also in the egg when the cortical granules are extruded in response to the stimulus given by the fertilizing spermatozoon. He was able to show that the ionophore A23187 (an agent considered to evoke a rise in intracellular calcium concentration) induces an ultrastructurally normal reaction in guinea pig spermatozoa, in the presence of free calcium in the medium. Since the reaction induced in this way occurred also at the same rate in the presence of several powerful protease inhibitors (Fig. 6.6),

Fig. 6.5. Graph showing how increasing numbers of motile guinea-pig spermatozoa react on addition of 2–4 mM $CaCl_2$, as a function of the time of incubation *in vitro* in a $Ca^{2+}$-free capacitation medium. The broken line shows the mean percentage of motile spermatozoa before addition of $Ca^{2+}$ and the solid line shows the mean percentage of spermatozoa undergoing activation and reacting within 10 minutes of addition of $Ca^{2+}$. (Fig. 3 in R. Yanagimachi and N. Usui, *Exp. Cell Res.* **89**, 161 (1974).)

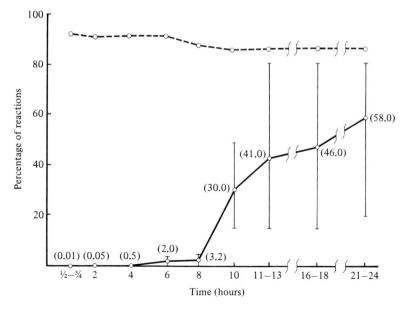

one must agree with his conclusion that it is probably the calcium ion as such, and not the action of acrosomal enzymes activated by it, that is immediately responsible for the events of fusion. The fact that calmodulin, a calcium-dependent enzyme activator, appears to be located in the acrosomal region makes this protein worth considering as the functional link through which the action of calcium is mediated within the membranes.

*Time relations of the reaction.* A difficult issue that is not yet resolved to everyone's satisfaction concerns the particular stage, prior to zona penetration, at which the acrosome reaction usually occurs in fertilizing spermatozoa, and so by inference the real function of the reaction. In the great majority of eutherian mammals the ovulated oocyte can be fertilized at any time over a period of 8–12 hours or sometimes rather longer. The cumulus and corona radiata of the unfertilized egg are lost progressively in the Fallopian tube during that time, and spermatozoa approaching several hours after ovulation may be presented with an almost naked oocyte, which many experiments show is quite penetrable in that state. In a normal setting, however, spermatozoa almost invariably are already available at the site of fertilization when oocytes arrive there, and the

Fig. 6.6. Graph demonstrating that even in the presence of 2 mM $Ca^{2+}$, natural protease inhibitors at 1 mg/ml (●, soyabean trypsin inhibitor; ○, lima bean trypsin inhibitor; □, ovomucoid) have no significant suppressive effect, as compared with the control medium (△), on the ability of the ionophore A23187 to induce acrosome reaction and loss in guinea pig spermatozoa. (Fig. 4 in D. P. L. Green, *J. Cell Sci.* **32**, 153 (1978).)

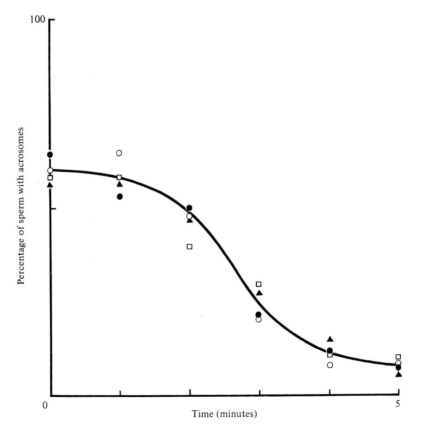

fertilizing spermatozoon penetrates the egg often before significant disruption of the cumulus has occurred. It is in fact very important for the normal development of the embryo that fertilization occur early, and delays of only a few hours seem to compromise the embryo's prospects seriously. In the physiological situation, therefore, the fertilizing spermatozoon will generally need to overcome an intact cumulus oophorus. The few studies extant suggest that capacitation is a requisite for penetration of the cumulus oophorus and that reacted spermatozoa are seen to move with relative ease through its matrix. Taken together, present evidence favours the view that spermatozoa must react at some initial point in their acquaintance with the intact cumulus to be able to penetrate it, and the ideas about the significance of capacitation and the acrosome reaction discussed earlier discount the possibility that the reaction in the fertilizing spermatozoon is necessarily induced by association with the zona surface.

The cumulus oophorus is usually shed from the fertilized egg well before syngamy (Fig. 6.1), and faster from the fertilized egg than the unfertilized egg. However, this rate can vary according to species. Cow and sheep eggs are reputed to lose the cumulus within a very few hours of ovulation whether or not fertilization occurs, and there are no cells at all around the opossum ovum at ovulation. The latter calls into question the postulated role of the cumulus for successful pick-up of the mammalian ovum from the ovarian surface and transfer to the tubal ampulla. Musk shrew (*Suncus*) eggs, on the other hand, are surrounded by a tight ball of cells (Fig. 6.7)

Fig. 6.7. Cumulus oophorus (CO) surrounding the egg of *Suncus murinus* (the musk shrew) recovered from the oviduct about 3 hours after ovulation. The cumulus cannot be dispersed by hyaluronidase or trypsin, probably because its cells are associating one with another by specialized junction points. The zona pellucida (Zp) around the egg can be visualized within the cumulus mass. × 280.

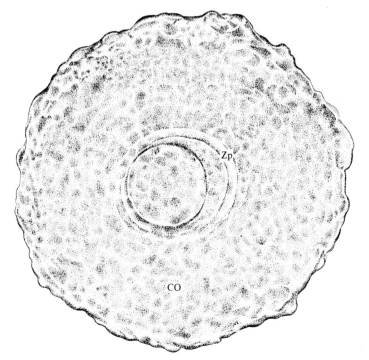

for up to 12 hours after fertilization, and since the cumulus is not sticky the eggs remain separate in the oviduct, unlike the diffuse cluster that soon forms there after ovulation from the separate masses surrounding the eggs of many rodents and rabbits. Oddly, the *Suncus* cumulus is completely resistant to hyaluronidase and several proteases that have been tried, probably because of the specialized junctions that form between its cells in the hours preceding ovulation, defying a precedent in other eutherian mammals where long-established junctions between periovular cells *disappear* shortly before ovulation. This strange arrangement in *Suncus* may have some relation to its giant acrosome (Fig. 6.8), and an understanding of the reasons for these special features might well help to create a more exact picture of what we regard as the normal sperm–egg interaction exemplified by the rabbit, rat, or indeed man.

It must be acknowledged finally that investigation has focussed very largely on the enzymatic content of the eutherian acrosome, without much evolutionary perspective on its composition in other vertebrate lines. It might be easier ultimately to assign specific functions for acrosomal enzymes if more were known of the acrosomal content in an evolutionary context, and observations of the spermatozoa in the marsupial *Didelphis* may in fact illustrate some of the pitfalls in assumption *a priori* of function for an acrosomal enzyme. Our recent investigations have established that the oocyte of the opossum loses all of its granulosa cell investment some

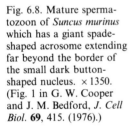

Fig. 6.8. Mature spermatozoon of *Suncus murinus* which has a giant spade-shaped acrosome extending far beyond the border of the small dark button-shaped nucleus. × 1350. (Fig. 1 in G. W. Cooper and J. M. Bedford, *J. Cell Biol.* **69**, 415. (1976).)

hours before ovulation and the spermatozoon needs negotiate only the naked zona pellucida. Yet, as well as acrosin and a comparable level of hyaluronidase, the opossum acrosome has some 40 times more aryl-sulphatase and up to 350 times the activity of *N*-acetylhexosaminidase than the rabbit acrosome. One must wonder how the function of these enzymes might have been interpreted had marsupials and not eutherians been the model for mammalian fertilization research.

## Interactions of spermatozoa with the zona pellucida
### Sperm–zona binding

Binding between the zona surface and specific elements on the plasma-lemma of the capacitated spermatozoon can also be a calcium-dependent event, and seems an essential preliminary to sperm penetration through the zona. We can reasonably assume that binding requires the presence of receptors on the plasma membrane, whether the acrosome has reacted or not, but we cannot say definitely yet just what sperm structures are essential for binding. However, the spermatozoon that has lost the outer acrosomal shroud beforehand (Fig. 6.3*c*) may subsequently be unable to establish a functional attachment at the zona surface, followed by penetration. The main problem to be attacked at the present time is the nature of the receptors on the sperm plasma membrane (in the region over the acrosome) and the complementary elements on the zona surface. Recently, Bonnie Dunbar has shown that the solubilized zona of pig and rabbit eggs contain three major glycoproteins, and Paul Wassarman and his colleagues have identified one as a sperm-binding element in mouse oocytes, which has a molecular weight of 83 000 and which they term ZP3.

### Passage through the zona

The function of the specific binding step has not been clearly delineated, though we can reasonably assume that the ligands provide a firm base from which the spermatozoon intrudes itself into the zona substance. It is quite clear that the reacted acrosomal shroud then is left behind at the zona surface and often remains bound to it. The spermatozoon generally cleaves a narrow path in the zona (Fig. 6.9), and light and electron microscope pictures show that it usually penetrates at about 45° to the tangent of the egg surface (though the penetration path is often curved). But the angle of approach varies ultimately on either side of this, such that occasional spermatozoa may penetrate the zona vertically or almost horizontally.

Spermatozoa have been observed to pass through the zona pellucida, of for instance the hamster egg, in 5–10 minutes; the means by which this is achieved is a vexed question. As well as the essential certainty that the sperm tail provides a necessary motive force, the most favoured explanation still attributes a major role to acrosin, the functional residue of which is thought to persist at the surface of the now-exposed inner membrane of the acrosome. Indeed, acrosin has often been called the 'zona penetration

Figs. 6.9. (*a*) Electron micrograph of the zona pellucida of a rabbit ovum showing a characteristic slit made by a spermatozoon in penetrating it. One can see several sections of the tail (T) of the spermatozoon responsible. PVS, perivitelline space between the zona and the egg itself. × 22 500. (*b*) Oblique section of the head of a rabbit spermatozoon (the tail is out of the plane of section) cleaving its way through the zona substance and so creating a penetration slit behind it. × 30 000. (Fig. 7 in J. M. Bedford, *Biol. Reprod., Supplement* **2**, 128 (1970).)

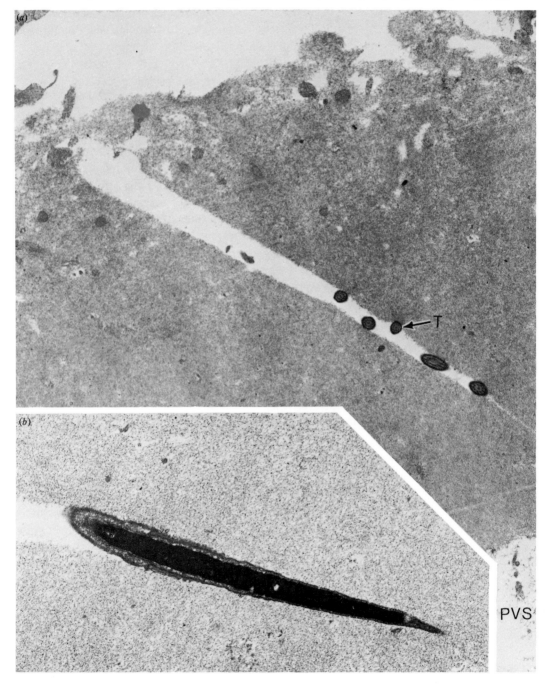

enzyme' though it is also possible that other enzymic residues might act cooperatively in this regard. The primary evidence for the idea that enzymic dissociation of zona substance in the path of the spermatozoon is the mechanism of penetration, is based on early observations of eventual distortion and sometimes dissolution of the zona pellucida by acrosomal extracts or mixed proteases from other sources, and also the apparent tendency of trypsin inhibitors to suppress fertilization *in vitro*, or sometimes when mixed with spermatozoa before insemination. I became concerned about the concept that the spermatozoon digests its way through the zona and questioned it in summarizing the first Gordon Conference on Fertilization in 1974, for reasons which now can be expanded. Ultrastructural studies clearly show the penetrating spermatozoon to be devoid of all visible content of the acrosome as it begins to enter and continues on its way through the zona (Fig. 6.9*b*). Although acrosin residues are measurable biochemically in spermatozoa from which the outer acrosome membrane has been removed artificially, no labelled acrosomal glycoproteins can be detected by autoradiography over the acrosomal region of spermatozoa penetrating the zona, nor do ultrastructural affinity markers for acrosin bind to the inner acrosomal membrane once the outer remnant has been discarded. Moreover, a significant increase in resistance of the (rabbit) zona to protease digestion does not necessarily have an influence on the rate of sperm penetration through it; and, while trypsin/acrosin inhibitors seem to inhibit binding to the mouse zona *in vitro*, they do not significantly

Fig. 6.10. This figure allows a direct comparison of the relative thickness (arrows) and consistency of the zona pellucida (Zp) in a marsupial (*Didelphis*) and a eutherian mammal (rabbit), whose eggs are approximately the same size. × 19 500.

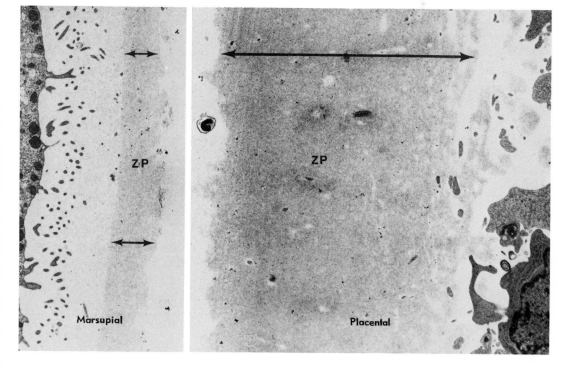

affect the ability of bound spermatozoa to penetrate it *in vitro*. Most telling, perhaps, is the fact that while highly purified trypsin may affect the biochemistry of some zona components it sometimes has little visible effect on the structure of the zona pellucida, and in some cases purified acrosin may not either.

In working with marsupial gametes, we have found major differences that bear further on this question. The opossum zona is dissolved by extracts of its spermatozoa within a few minutes, a 0.1% solution of purified commercial trypsin does so in only 2–3 seconds, and penetration of the spermatozoon creates a large gap in this marsupial zona that is compatible with the notion of a process of zona digestion. Its flimsy character contrasts with the rabbit or pig zona of approximately similar circumference that has a physical resilience to deformation, is denser and thicker (Fig. 6.10), is relatively resistant to proteases, and in which the spermatozoon cleaves a narrow slit, really no greater than the dimensions of the sperm head. This evidence is very suggestive that enzymic digestion may have a major place in enabling the spermatozoon to penetrate the zona pellucida of the marsupial egg, but it is difficult for these and other reasons discussed later to avoid the impression that physical forces have come to play an additional and major part in sperm passage across the zona pellucida in eutherian mammals.

**Gamete fusion and sperm incorporation by the egg**
*Mechanisms*
One of the surprises arising from ultrastructural investigation in the late 1960s was the complexity of the way the fertilizing spermatozoon is incorporated by the egg in eutherian mammals (Fig. 6.11). After successfully breaching the zona pellucida, the spermatozoon may swim for a brief period in the perivitelline space, and stops at about the time it clearly becomes anchored to the egg surface. Electron microscope studies give unequivocal proof that fusion with the egg is always established by way of the mid-region of the sperm head. The most recent studies make it clear also now that the equatorial segment is intact in perivitelline spermatozoa, that the fertilizing spermatozoon then apparently uses the restricted region of plasma membrane overlying the persistent equatorial segment of the acrosome to accomplish fusion (Fig. 6.11*b*), and that the segment remains intact after the spermatozoon has been incorporated into the ooplasm (Fig. 6.13). This gives a clue to the possible significance of this quiescent specialized posterior region of the acrosome. Its particular stability, probably created in part by structural protein links between its parallel membranes, excludes it from the fusion events of the acrosome reaction, so preserving intact a segment of unencumbered sperm plasma membrane that is potentially fusogenic. A propensity for fusion with the oolemma is not exhibited until the acrosome reaction has occurred, implying that

Fig. 6.11. Sequence of events involved in fusion and incorporation of the fertilizing spermatozoon by the mammalian egg. After penetrating the zona pellucida, the spermatozoon in the perivitelline space (*a*) uses the surface overlying the equatorial segment (delineated by arrowheads) to establish fusion (*b*) with the vitelline surface. Fusion stimulates release of cortical granules, decondensation of the sperm nucleus begins, and soon thereafter (*c*) the egg cortex engulfs the front of the sperm head bounded by the inner membrane of the acrosome. When this is completed (*d*), that region of the sperm nucleus is encased by a vesicle composed of internalized egg membrane (EM), persisting equatorial segment (delineated by arrowheads, and see Fig. 6–13) and inner acrosomal membrane (IA). CG, cortical granule; MV, microvillus. (After Fig. 22 in J. M. Bedford and G. W. Cooper, *Cell Surface Reviews*, vol. 5, *Membrane Fusion*, pp. 66–125. Ed. G. Poste and G. L. Nicolson. Elsevier/North Holland Biomedical Press; Amsterdam (1978).)

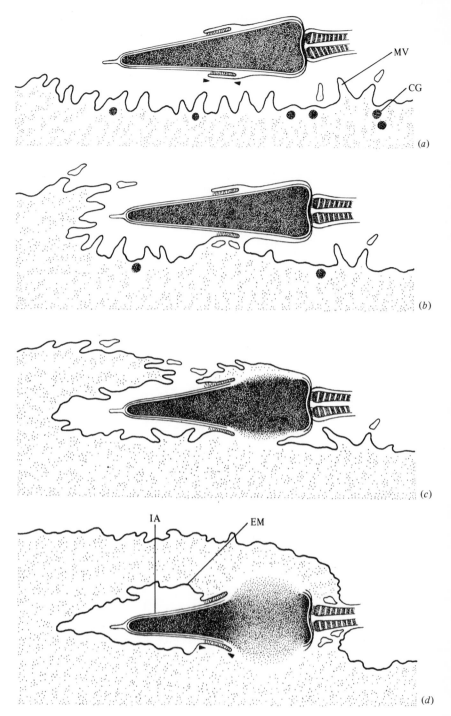

Fig. 6.12. Scanning electron micrograph of an early stage of sperm fusion *in vivo* (*b*, *c* in Fig. 6.11) with a hamster egg (inset) from which the zona pellucida was removed after fixation. Microvilli are present over the whole egg surface except one small region (inset) that overlies the metaphase spindle, and they also cover the mid-region of the sperm head (the approximate location of the equatorial segment is outlined in Fig. 6.16), leaving the region overlain by inner acrosomal membrane (IA) and the postacrosomal region (PA) still uninvolved with the egg surface. × 13000; *Inset*, × 750. (Courtesy of Dr David Phillips.)

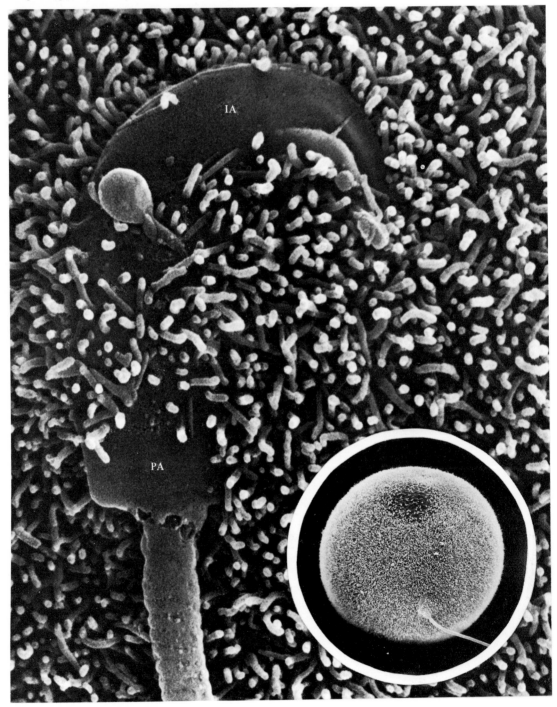

the reaction propagates or confers on the specialized region overlying the equatorial segment some change appropriate to its fusion role.

Although the precise event of fusion is not clearly defined, occasional views of early stages obtained with the scanning electron microscope such as that shown in Fig. 6.12 suggest that it may involve initial interaction with egg microvilli. Nevertheless relatively little is known about the molecular mechanism involved in fusion between the spermatozoa and the egg plasma membrane or oolemma, in mammals. As noted below, the zona pellucida constitutes the major specific barrier and there is much less specificity to the subsequent event of fusion. Calcium again seems to be an essential element for fusion to occur and, in the hamster at least, fusion begins to be suppressed when the temperature falls below about 25 °C. The oolemma receptor system may vary somewhat from mammal to mammal. Trypsin is believed to reduce the receptivity of the mouse oolemma, whereas several different proteases, glycosidases, and a variety of lipases have no effect on sperm fusion with the zona-less hamster egg. The only enzyme that does affect this, and powerfully too, is phospholipase C, indicating a critical role for the molecular order of oocyte membrane lipids in the fusion process. There is no comparable information for other species, but the indifference of the hamster oolemma to such a variety of enzymes, excepting phospholipase C, may be significant in relation to the fact that the hamster egg has been found to be receptive to the spermatozoa of almost all other mammals that have been tested.

Fusion is followed within a few minutes by several further events. First, it triggers the general sequence of egg activation, including the extrusion, or exocytosis, of cortical granules and also formation of the second polar body, discussed below. There does not appear to be an elecrical block to polyspermy as in some invertebrate eggs, but experiments on hamster eggs suggest that a fusing spermatozoon induces a hyperpolarization that is probably a recurring event independent of further stimulation by spermatozoa. These hyperpolarizations are due to an increase in potassium permeability caused by a rise in intracellular calcium that accompanies activation.

Soon after fusion the sperm nuclear envelope disappears, the ooplasm appears to insert into the post-nuclear region of the sperm head and this is followed by an engulfment of the foremost or rostral portion of the sperm head (Fig. 6.11). As a consequence, the front of the fertilizing spermatozoon is incorporated within an envelope of membrane, most of the outer leaf of which is egg membrane and the remainder acrosomal in origin (Fig. 6.13). It is possible that sperm plasma membrane overlying the posterior head region and tail of the spermatozoon remains as part of the egg surface, temporarily at least. There is no certain proof of this yet, but Mike O'Rand has shown that antibody prepared against an integral sperm membrane protein can damage fertilized one-cell eggs, but not unfertilized or parthenogenetically activated eggs, and this implies that sperm membrane components do become part of the fertilized egg surface.

The pattern of sperm–egg fusion just described seems to be unique to mammals, for observations on the lamprey, salamander and chicken suggest that their spermatozoa fertilize as do invertebrate spermatozoa; they fuse with the oolemma by the *apex* of the inner acrosomal membrane (commonly forming the surface of a protruded 'acrosome filament' in marine invertebrates), and the head enters the ooplasm devoid of mem-

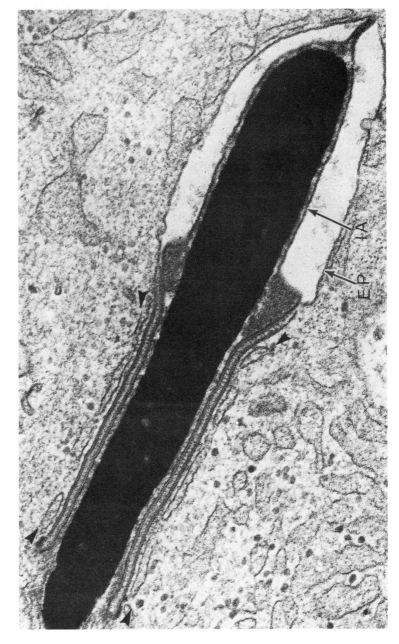

Fig. 6.13. The nucleus of the fertilizing hamster spermatozoon is surrounded, as depicted in Fig. 6.11d, by egg plasma membrane (EP), inner acrosomal membrane (IA) and an intact equatorial segment (delineated by arrowheads). Note the shoulders of subacrosomal material (see Fig. 6.17) positioned immediately in front of the equatorial segment. × 57000. (Fig. 6 in H. D. M. Moore and J. M. Bedford, *J. Ultrastr. Res.* **63**, 110–17. (1978).)

brane. The form and structure of the monotreme spermatozoon suggests that it probably fertilizes in that way, and recently John Rodger and I have been able to establish that marsupial (*Didelphis*) spermatozoa associate with the oolemma by the inner acrosomal membrane and unlike those of eutherian mammals carry no membrane around the head into the ooplasm (cf. Figs. 6.13 and 6.14). Thus, marsupial spermatozoa follow what has been the 'party line' among both the invertebrate and other non-mammalian vertebrate species that have been studied in any detail. More

Fig. 6.14. Transmission electron micrograph of a fertilizing spermatozoon soon after entering a marsupial (*Didelphis*) egg. By contrast with Fig. 6.13, the decondensing nucleus (outlined by small arrowheads) is quite free of any investing membrane, as in sub-mammalian species, and the tail (T) has become dissociated from it. × 14800. (Fig. 10 in J. C. Rodger and J. M. Bedford, *J. Reprod. Fert.* **64**, 171 (1982).

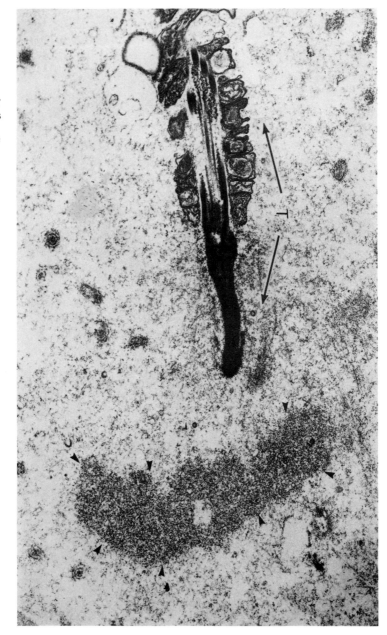

evidence is needed for other vertebrates, but that available now does suggest that eutherian mammals practise a mode of gamete interaction peculiarly their own.

### Evolutionary significance

In considering the reasons for this abrupt and radical departure to a unique mode of sperm incorporation by eutherian mammal eggs, it may be useful to examine briefly the way both male and female gametes have changed during the transition from our reptile-like ancestors to the eutherian state. The simple filamentous form of the monotreme sperm head (Fig. 6.15, *left*), unresistant to even weak physical and chemical forces, has been replaced

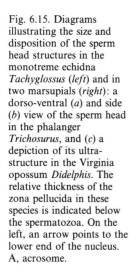
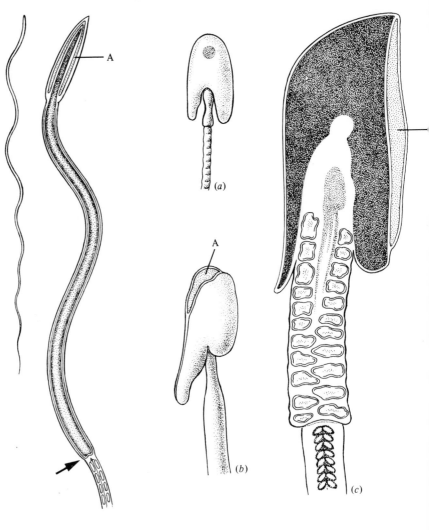

Fig. 6.15. Diagrams illustrating the size and disposition of the sperm head structures in the monotreme echidna *Tachyglossus* (*left*) and in two marsupials (*right*): a dorso-ventral (*a*) and side (*b*) view of the sperm head in the phalanger *Trichosurus*, and (*c*) a depiction of its ultrastructure in the Virginia opossum *Didelphis*. The relative thickness of the zona pellucida in these species is indicated below the spermatozoa. On the left, an arrow points to the lower end of the nucleus. A, acrosome.

in eutherian mammals by a more compact, strikingly keratinoid head in which the nucleus and peri-nuclear material, including the perforatorium, are made rigid by extensive disulphide cross-links, and the inner acrosomal membrane also displays a distinctive resilience (Fig. 6.16). In addition, the head has developed the stable posterior equatorial region of the acrosome (equatorial segment) which is apparently unique to eutherian spermatozoa. Coincidentally, the vestment of the egg has been transformed from the thin frail zona of the monotreme ovum to the broad elastic structure that

Fig. 6.16. Sagittal section of a generalized eutherian sperm head that has lost the reacted elements of the acrosome. This illustrates features not found in other mammals (cf. Fig. 6.15) but which have appeared *de novo* in the sperm head in all eutherians. Diagrams representing rabbit (bull, boar or ram) (above left) and hamster (below left) show the disposition of the equatorial segment in the sperm head. The relative thickness of the zona pellucida of the eggs of eutherian mammals such as rabbit, man, pig, cow can be compared with the zona in a monotreme and marsupials (Fig. 6.15) (refer also to Fig. 6.10).

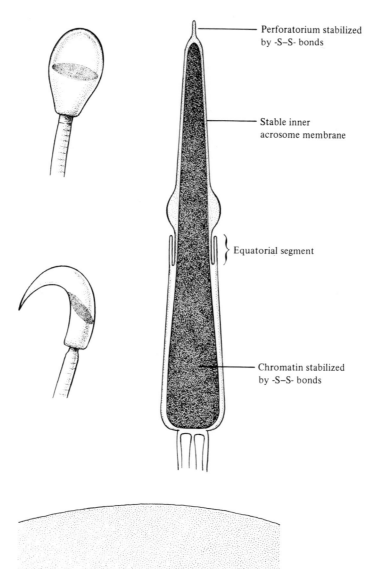

Perforatorium stabilized
by -S–S- bonds

Stable inner
acrosome membrane

Equatorial segment

Chromatin stabilized
by -S–S- bonds

typifies the eutherian zona (Fig. 6.16). In each case, marsupial gametes lie somewhere in between (Fig. 6.15). The marsupial sperm head has become more compact but with no special stability in the nucleus or in its head membranes. The marsupial zona, though somewhat more prominent than that around monotreme eggs (Fig. 6.15), still represents a minimal barrier in a physical sense (Fig. 6.10).

While we can only speculate, these observations together suggest that ultimately it may have been the increasingly formidable egg vestments that have evoked adaptations in the sperm head. For example, if sperm passage through the eutherian zona is largely a physical event involving shear forces at the inner acrosomal membrane surface, a stability required to withstand these forces may have been incompatible with fusogenic properties in that membrane, thus necessitating the special adaptation of a different fusogenic region – the equatorial segment – found so far only in eutherian spermatozoa. Though not obvious in all eutherian mammals, it is intriguing that several at least seem to have developed an arrangement that would tend to protect the putatively less stable portion of the plasmalemma over the equatorial segment. In some species, the cross-linked peri-nuclear material is gathered as a prominent shoulder extending across the sperm head immediately rostral to the segment (Fig. 6.17) in a fashion that would locally divert the zona substance and so minimize its interaction with the surface over the equatorial segment. The alternative arrangement in which the segment tends to rest in a shallow depression in the contour of the nucleus (Fig. 6.17) would serve the same end. Although these nuances of

Fig. 6.17. Diagram of the arrangements of the mid-region of two sperm heads supposedly in the process of penetrating the zona pellucida from upper left to lower right. On the left a shoulder of subacrosomal material diverts the zona substance, minimizing its interaction with the membrane overlying the equatorial segment. On the right the position of that segment (arrowheads) in a shallow depression in the nucleus achieves the same end. Arrows highlight the interface of friction between the inner membrane of the acrosome and the substance of the zona.

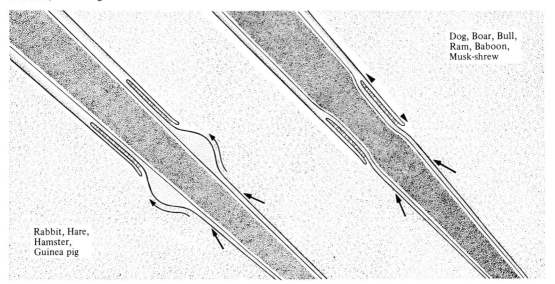

Dog, Boar, Bull,
Ram, Baboon,
Musk-shrew

Rabbit, Hare,
Hamster,
Guinea pig

sperm structure might be interpreted as adaptations that are a response to the resilient dense nature of the zona substance, it is more difficult to try to explain why the eutherian ovum should have invested itself with its distinctive zona pellucida.

## Activation of the egg

Fusion appears soon to inhibit further movement of the fertilizing spermatozoon, and it does several things to the egg. As noted earlier, electrical changes occur in the egg surface and a resting potential of about $-30$ mV (hamster) is increased for a few seconds to about $-75$ mV, finally remaining at about $-40$ mV in the fertilized egg, though subsequent hyperpolarizations may occur then without a further fusion stimulus. Fusion also evokes exocytosis of cortical granules within a few minutes, and it is their content that effects a change in zona pellucida, and/or oolemma according to species, which is responsible for the block to polyspermy. Cortical granule exocytosis is only a consequence of, and not in fact an essential requirement, for activation, since this often fails to occur in artificially activated oocytes. Fusion also stimulates the metaphase spindle of the secondary oocyte to complete its division, one half of the chromatin then being extruded within the second polar body (Fig. 6.1), an event that leaves the oocyte in the haploid state. Expression of the activated state almost certainly brings many other as yet unrecognized changes in the state of the ooplasm linked to a rise in intracellular calcium. Protein synthesis probably utilizes stored messages or information already present in the oocyte at ovulation, since transcription of new RNA does not occur until about the 2- to 4-cell stage in the developing embryo.

Activation does not necessarily require the stimulus of the spermatozoon, and can be induced for example by temperature shock or by pricking of the egg membrane. Activation without a spermatozoon – parthenogenesis – often leads to the beginnings of development as far as the blastocyst stage, and in certain vertebrates, a good example being the turkey, advanced development can result; it is very limited in mammals (see Chapter 3).

### Pronucleus formation and syngamy

The mechanisms of pronucleus formation have been slow to emerge because their elucidation must depend on indirect inference rather than direct observation. The factors that transform the female elements probably have many features in common with those that transform the sperm head. Nonetheless, occasional observation of anomalous situations shows that the female pronucleus can form normally where the sperm head has remained condensed, suggesting that there are some differences in this respect. Relatively little is known yet of the details of female pronucleus formation. The chromatin remaining after second polar body extrusion becomes diffuse and presumably is subjected to a modification of the associated basic proteins as it is transformed into the female pronucleus surrounded by a nuclear envelope that appears *de novo*.

When the sperm head enters the ooplasm, the nuclear envelope disperses immediately and decondensation of nuclear material begins in the exposed mid-posterior region. Swelling of the dense sperm chromatin has been assumed to require efficient reduction within it of disulphide cross-links and, as the sperm nucleus enlarges as a mesh of nucleoprotein strands, basic cysteine-rich protamine that has also a high content of arginine disappears from its close association with the sperm DNA. This was first hinted at in the observation that radiolabel disappears coincidentally with the decondensation phase in the heads of arginine-labelled spermatozoa. Somatic histones associate with the sperm's DNA only as this becomes recognizable as a pronucleus. Thus, there seems to be a period after decondensation and before final formation of the pronucleus when the DNA of the sperm head is in a transitional state in which the character of the proteins associated with it undergoes radical change. Not surprisingly, replication of DNA occurs between 5 and 8 hours after sperm penetration while it is in this relatively non-repressed state.

The expanded sperm chromatin becomes reconstituted as a pronucleus, first identifiable as such by the reformation of a nuclear envelope as a series of flattened vesicles, followed later by the appearance of a variable number of dense nucleoli. The male pronucleus is the more active mover in the mutual approach of the formed pronuclei that occurs eight or more hours after gamete fusion, according to species. The final event, syngamy, does not involve fusion (as in invertebrates) but rather a regional random breakdown of the envelopes of the apposed pronuclei. Then, as the diffuse chromatin becomes organized into distinctive chromosomes, newly synthesized microtubules pass through the gaps in the nuclear envelope as precursors to the first cleavage spindle, the appearance of which marks the end of syngamy, and this is followed soon by the first cleavage division.

*The block to polyspermy*

The whole problem of ensuring that only one spermatozoon contributes genetic material to the embryo is dealt with in a variety of ways by different animals. Some allow several spermatozoa to enter the ooplasm and then exclude all but one from participation in syngamy (a stratagem used among birds and reptiles). Fertilization appears typically monospermic in mammals and to achieve this they erect physiological barriers that block the entry into the ooplasm of all except the first fertilizing spermatozoon. The speed of this block is critical for species with external fertilization, where a large number of fertile spermatozoa are deposited in the immediate vicinity of the egg, but it is not so urgent for mammals since the regulatory mechanisms of the Fallopian tube allow the passage of only very small numbers of spermatozoa to the site of fertilization (see Chapter 5). The importance of this regulation of sperm numbers in mammals for avoidance of polyspermy is seen particularly in fertilization *in vitro* where the incidence of polyspermy is often elevated as a consequence of the abnormally high number of spermatozoa that can contact the unfertilized egg.

Nonetheless, even in mammals the most important defence against polyspermy is mounted by the egg itself. The ability to establish this block is another feature that makes its appearance as a part of the final phase of maturation, between resumption of meiosis and ovulation, as discussed earlier; the primary oocyte plucked from the follicle does not respond to a fertilizing spermatozoon by exocytosis, and so polyspermy is common when such oocytes are placed in the Fallopian tube. The presence of relatively few spermatozoa – perhaps a few hundred at most – at the fertilization site in mammals, renders unnecessary a fast or electrical block of the type seen in sea urchin eggs; the mammalian block apparently depends solely on exocytosis of cortical granules stimulated by the fertilizing spermatozoon (Fig. 3.3).

Even among mammals the precise mechanism of the block varies according to species. Exocytosis in the rabbit oocyte brings no functional alteration to the zona pellucida, but is known to result in a change in the charge density of the surface of the oolemma and in its reduced receptivity for spermatozoa (shown dramatically by the many freely swimming supplementary spermatozoa that crowd the perivitelline space – Fig. 3.7). Whether this oolemmal block reflects a removal of or conformational change in sperm receptors by enzymes from cortical granules or addition of masking elements is difficult to say. The converse holds for many other species, including the common laboratory rodents: the oolemma often remains quite receptive to a second spermatozoon and the most important action of the content of the cortical granules appears to involve modification of an essential sperm receptor at the zona surface. This receptor, identified in the mouse as a glycoprotein ZP3, can no longer be distinguished as such in the zona of the fertilized egg. Fertilization, or cortical exocytosis induced by a calcium ionophore, brings structural change in a second major glycoprotein of the mouse zona, ZP2, the functional significance of which is not yet clear.

Exocytosis also results in a physical change in the properties of the zona that has been termed 'hardening' and which occurs some 2–3 hours after activation: this is shown by an increased resistance to protease digestion and may result from cross-linking of tyrosine residues in the zona substance induced by an ovoperoxidase of cortical granule origin. Whether this enzyme-induced hardening plays a significant part in the zona block has not been established. Hardening occurs in the rabbit zona, yet does not seem to affect sperm passage through it. We can surmise that in species in which the zona is the critical element in the prevention of polyspermy, modification of zona surface receptors could be the most important way an effective block is established.

**Specificity of fertilization**

A specificity of complementary 'receptors' on the surface of the gametes themselves can play an important part in the maintenance of a reproductive isolation between species. Though fertilization may occur between some closely related animals, e.g. sheep and goat, rabbit and hare, ferret and mink, this restriction applies among most mammals, and cross-fertilization even between species within the same Family is normally not possible. Among the preliminaries, specificity is often expressed also in the failure of foreign spermatozoa to be transported to the site of fertilization, and in many though by no means all instances, foreign spermatozoa survive in the female tract less well than do native spermatozoa. However, where they do survive, spermatozoa may become capacitated and fertilize eggs of their own species in a foreign oviduct. Squirrel monkey spermatozoa will fertilize squirrel monkey eggs in the rabbit oviduct, and rabbit spermatozoa fertilize rabbit eggs in the mouse or hamster oviduct. On the other hand, a capacitated spermatozoon generally cannot penetrate a foreign zona pellucida. We are not in a position yet to explain the precise mechanism of that specificity, but it seems to lie at least partly in the inability of capacitated spermatozoa to establish a firm binding relationship with the foreign zona surface. Even though they may not penetrate a foreign zona, non-capacitated spermatozoa of many mammals frequently will attach non-specifically to its surface, but the attachment affinity seems to become more specific after capacitation. Human spermatozoa, however, capacitated or not, fail to display even a non-specific affinity for the surface of the zona of non-hominoid primates and other mammals, but they bind to and will penetrate the zona of the gibbon *Hylobates lar*, the ape furthest from man, and one might assume that they are compatible with eggs of the other true apes, the gorilla, orangutan, and chimpanzee.

The pre-eminence of the zona over the oolemma as the prime controller of interspecific fertilization is very evident when we study the wider options for cross-fertilization that can follow removal of the zona. A degree of specificity does persist even for zona-less oocytes, but the rules that order this are not evident and the situation in each case must be established by experiment. Perhaps the most interesting and now most useful system in this respect is the hamster egg, which seems to hold open house at the level of the oolemma – Ryuzo Yanagimachi and his colleagues have shown that, unlike rat or mouse eggs, the zona-less hamster egg will fuse with and incorporate human spermatozoa.

With the occasional exception, there is little or no specificity to the next step, the interaction between the sperm chromatin and the ooplasmic factors needed for its transformation, and the fact that hamster ooplasm will transform sperm chromatin to a pronucleus has been adapted to bring to expression the chromosomal constitution of human spermatozoa. This can be achieved in oocytes treated with colcemid which prevents the organization of chromosomes on a spindle. When it is refined further, the procedure could have important application in the screening for sperm chromosomal defects in man.

**Errors of fertilization**

The fine tuning by hormones of the timing of insemination and ovulation, and the egg's intrinsic controls of the steps that lead to syngamy, all ensure that fertilization will result in a normal healthy fetus. Occasionally, however, errors of fertilization occur that bring abnormal chromosome numbers in the egg. *Polyspermy* (fertilization by more than one spermatozoon) is seen naturally in a very low percentage of cases, but can be enhanced by ageing of the egg before fertilization. *Polygyny*, also a rather rare error, follows from a failure to extrude the second polar body resulting in an extra set of female chromosomes. Very uncommonly, *gynogenesis* takes place, as a result of failure of the male pronucleus. By contrast, the contribution by the spermatozoon of all the genetic material, *androgenesis*, we now know to be the cause of hydatidiform mole. The ways in which this and other errors of fertilization result in chromosomal defects in the offspring is discussed more fully in Book 2, Chapter 5.

The coincident ageing of the egg before ovulation that is an unavoidable consequence of ageing of the mother may bring an increased incidence of defects due to errors of meiosis (see Chapter 2), particularly non-disjunction. Such a failure results in an extra chromosome in the female pronucleus that should have passed to the first polar body before ovulation. Fertilization of such an egg then creates the condition of *trisomy* – an important example is the extra chromosome 21 seen in Down's syndrome in man, where development often continues to birth and the condition is expressed later in the abnormal appearance and mental development of the child.

In contrast to the ageing that is an inevitable part of the ageing of the female, pre-fertilization ageing of gametes over the relatively short period *after* their release from the gonads greatly increases the chances of lethal chromosomal abnormalities prejudicial to development of the embryo beyond the implantation stage. In nature, the female in most mammals below man generally becomes sexually receptive and then mates some hours before ovulation, or else, as in rabbits, ferrets and shrews, ovulation is reflexly induced by the act of copulation. This means that capacitated spermatozoa are waiting for eggs as they are ovulated, an arrangement that makes good sense since it diminishes the likelihood that such pre-fertilization ageing of spermatozoa or eggs can take place. But gamete ageing is apt to be important in man, in whom lack of such a synchrony means that spermatozoa may sometimes not fertilize until perhaps 40–60 hours after insemination, or with insemination after ovulation, the egg will age before fertilization. There is little doubt that ageing of ova for only a few hours before fertilization does indeed compromise the chance of their developing as normal embryos (see Chapter 3), and such a delay is often associated with embryonic death and embryonic chromosomal anomalies such as triploidy or mixoploidy. The issue is not so clear for the spermatozoon.

In considering lethal factors that arise in spermatozoa after their release from the testis, Marcus Bishop concluded in 1964 that mutations in the sperm nucleus are likely to increase in frequency as a function of time, and it is a common view today that ageing brings an increase in genetic defects in spermatozoa. Damage can be induced in sperm DNA alone, and selective damage to the eutherian sperm genome without attenuation of the ability to fertilize does occur after their exposure to sublethal doses of alkylating agents or to ionizing radiation. Thus, the possibility for dissociation of the genetic and egg-penetrating functions of spermatozoa is very real. On the other hand, from a teleological standpoint, it is at least arguable that, where there is no specific insult other than ageing, the margin of safety built into the stable sperm nucleus in eutherian mammals is unlikely to be so narrow that ageing could compromise its integrity before that of the ability to fertilize. In fact, there is yet no real evidence of an increase of point mutations or specific age-related change in eutherian sperm DNA. It is important to realize, moreover, that insemination with ageing spermatozoa carries a very real risk of *delayed fertilization* because of low numbers of competent spermatozoa ultimately at the site of fertilization. Thus, the chromosomal anomalies observed in early embryos after insemination of ageing spermatozoa, which are similar in their variety to those that follow delayed fertilization, could very well really reflect ageing of the ovum and not the sperm nucleus. In dwelling on this point, one must reflect that although there is no indication of any fundamental difference in their make-up, the spermatozoa of certain bats especially, and to a lesser degree hares, seem able to undergo a prolonged ageing of weeks in the female with no deleterious effects on their later ability to support the development of viable embryos.

**Fertilization *in vitro***

For many years there was ample reason to feel uncertainty about claims that mammalian eggs had been fertilized in culture conditions outside the body. The first successful experiments of this kind were made by Charles Thibault and Louis Dauzier in 1954 with rabbit gametes, but it was left to M. C. Chang to produce irrefutable evidence of fertilization *in vitro* in 1959 by going on to culture eggs *in vitro* until the 4-cell stage, transplanting them to the oviducts of a female, and later obtaining live young of both sexes. Those manipulations paid careful attention to various aspects such as the pH and gas content of the culture medium. The main trick, however, depended on an awareness of the functional immaturity still of epididymal and ejaculated spermatozoa, and the need to use capacitated spermatozoa – in this case flushed from the uterus of another female rabbit about 10 hours after mating. Since that demonstration, both capacitation and fertilization (verified either through transplantation of eggs or cytologically) have been accomplished successfully *in vitro* in many species that include all the common laboratory rodents, the cat, sheep, cow, dog and monkey, though not the pig. The most dramatic result by far was achieved through the persistence of Bob Edwards and Patrick Steptoe over a period of about ten years, during which they adapted the techniques to human gametes for collection of eggs, *in vitro* capacitation of spermatozoa, fertilization, and zygote transfer, and also determined the conditions necessary for implantation and the continuation of pregnancy in their subjects. This culminated in 1979 in the birth of a baby girl after fertilization *in vitro* of an egg taken close to the time of natural ovulation from the ovary of her mother, whose previous infertility was attributable to blocked Fallopian tubes.

Fertilization *in vitro* has other possible uses now. It is employed by many investigators to study different aspects of the physiology and morphology of the fertilization process. It is used also as an assay of the suitability of different experimental conditions to achieve a complete functional state of capacitation for the spermatozoa of a variety of species. However, herein lies a need for caution. The prolonged periods of 15–24 hours over which the gametes are sometimes cultured together means that, in some cases, spermatozoa have been able to complete their capacitation in the in-vitro fertilization medium before penetrating the eggs. This would give a wrong or exaggerated impression of the sperm capacitation potential of the conditions under test. Under these circumstances a wise precaution is that the sample be left for no more than 3–4 hours with eggs before these are assessed for penetration.

As we noted in the discussion of specificity, fertilization *in vitro* of zona-less hamster eggs seems to offer some promise now also for assessment of the ability of human spermatozoa to undergo an acrosome reaction and then to fuse with and be incorporated by an oocyte. Results beginning to emerge suggest that this will be useful in identifying certain men who produce motile spermatozoa with previously undetectable defects that prevent them from interacting appropriately with eggs. Several other elements of the conception process, e.g. transport through cervical mucus and further to the tubal ampulla, and also the ability to penetrate the cumulus oophorus and especially the zona pellucida, are of course not monitored by this test. Thus, penetration of a zona-less oocyte alone, and so a *positive* result in this in-vitro system, may not mean very much as a test for fertility. On the other hand, a consistent *negative*, i.e. a failure of visibly normal motile spermatozoa to penetrate, may be very useful for the diagnosis of subtle defects of sperm fertility not revealed by conventional evaluation.

As a keystone of sexual reproduction, fertilization retains in mammals the fundamental role of activating the egg and re-establishing the full diploid chromosome complement needed for the normal development of a new individual. Fertilization has generally been more difficult to study in mammals than in invertebrates in which both gametes can often be obtained in large numbers for analysis, and their reactions can be timed precisely. Nevertheless, it is no longer necessary or appropriate that fertilization be taught to students of mammalian biology primarily by reference to the classical invertebrate models. Over the last decade, fertilization in mammals has become much less of a mystery, and this has led to changes in some concepts about the mechanisms through which their gametes achieve syngamy.

Gamete organization in the primitive monotreme mammals has remained essentially similar to that of reptiles and many birds, and a similar mode of fertilization is likely. On the other hand, the spermatozoa of therian mammals (eutherians and marsupials) seem to have established a closer dependence on the male and female tract for their final maturation, and the eggs of eutherian mammals incorporate the fertilizing spermatozoon in a unique way that is radically different from that practised by the invertebrates and all other vertebrates that have been studied. Why this is so is not really clear, but some of this change may be due to parallel modifications in the character of the ovum, and there probably remain yet unrecognized even more subtle determinants of the conception process. Thoughtful focus also on other groups of higher vertebrates, especially those with gamete features of special interest, should, with the increasingly precise analytical techniques of cell biology, help solve many of the uncertainties that still exist for each of the principal steps in mammalian fertilization.

### Suggested further reading

Male reproductive systems. In *Handbook of Physiology, Section 7, vol. 5*. Ed. D. W. Hamilton and R. O. Greep. American Physiological Society; Washington, DC (1975).

*The Spermatozoon.* Ed. D. W. Fawcett and J. M. Bedford. Urban and Schwarzenberg; Baltimore (1979).

*Fertilization Mechanisms in Man and Mammals.* R. B. L. Gwatkin. Plenum Press; New York (1977).

The mechanism of the acrosome reaction. D. P. L. Green. In *Development in Mammals*, vol. 3, pp. 65–83. Ed. M. H. Johnson. North Holland; Amsterdam (1977).

Membrane fusion events in the fertilization of vertebrate eggs. J. M. Bedford and G. W. Cooper. In *Cell Surface Reviews*, vol. 5: *Membrane Fusion*, pp. 66–125. Ed. G. Poste and G. L. Nicolson. North Holland; Amsterdam (1978).

*Immunobiology of Gametes.* Ed. M. Edidin and M. H. Johnson. Cambridge University Press (1977).

*Conception in the Human Female.* R. G. Edwards. Academic Press; London and New York (1980).

*Bioregulators of Reproduction.* Ed. G. M. Jagiello and H. J. Vogel. Academic Press; London and New York (1981).

*Biology of Fertilization.* Ed. C. B. Metz and A. Monroy, Academic Press; London and New York (in press).

Mechanisms of fertilization in mammals. R. Yanagimachi. In *Fertilization and Embryonic Development* in vitro, pp. 81–182. Ed. L. Mastroianni and J. D. Biggers. Plenum Publishing; New York.

# INDEX